Claudio G. Rolli

In vitro model systems to study single and collective cell migration

Claudio G. Rolli

In vitro model systems to study single and collective cell migration

The Social Web in a Dish Solitary Riders and Crowd Communication

Südwestdeutscher Verlag für Hochschulschriften

Impressum / Imprint
Bibliografische Information der Deutschen Nationalbibliothek: Die Deutsche Nationalbibliothek verzeichnet diese Publikation in der Deutschen Nationalbibliografie; detaillierte bibliografische Daten sind im Internet über http://dnb.d-nb.de abrufbar.
Alle in diesem Buch genannten Marken und Produktnamen unterliegen warenzeichen-, marken- oder patentrechtlichem Schutz bzw. sind Warenzeichen oder eingetragene Warenzeichen der jeweiligen Inhaber. Die Wiedergabe von Marken, Produktnamen, Gebrauchsnamen, Handelsnamen, Warenbezeichnungen u.s.w. in diesem Werk berechtigt auch ohne besondere Kennzeichnung nicht zu der Annahme, dass solche Namen im Sinne der Warenzeichen- und Markenschutzgesetzgebung als frei zu betrachten wären und daher von jedermann benutzt werden dürften.

Bibliographic information published by the Deutsche Nationalbibliothek: The Deutsche Nationalbibliothek lists this publication in the Deutsche Nationalbibliografie; detailed bibliographic data are available in the Internet at http://dnb.d-nb.de.
Any brand names and product names mentioned in this book are subject to trademark, brand or patent protection and are trademarks or registered trademarks of their respective holders. The use of brand names, product names, common names, trade names, product descriptions etc. even without a particular marking in this works is in no way to be construed to mean that such names may be regarded as unrestricted in respect of trademark and brand protection legislation and could thus be used by anyone.

Coverbild / Cover image: www.ingimage.com

Verlag / Publisher:
Südwestdeutscher Verlag für Hochschulschriften
ist ein Imprint der / is a trademark of
OmniScriptum GmbH & Co. KG
Heinrich-Böcking-Str. 6-8, 66121 Saarbrücken, Deutschland / Germany
Email: info@svh-verlag.de

Herstellung: siehe letzte Seite /
Printed at: see last page
ISBN: 978-3-8381-3956-2

Zugl. / Approved by: Heidelberg, Ruprecht-Karls-Universität, Diss., 2011

Copyright © 2014 OmniScriptum GmbH & Co. KG
Alle Rechte vorbehalten. / All rights reserved. Saarbrücken 2014

The Social Web in a Dish.

Solitary Riders and Crowd Communication.

Contents

Contents	iii
Abstract	vii
Zusammenfassung	ix
Abbreviations	xi

I Introduction 1

1 Cell migration 3
 1.1 Model systems to study cell migration *in vivo* 4
 1.2 Model systems to study cell migration *in vitro* 4
 1.3 Surface functionalization to mimic the extracellular matrix 7
 1.4 Surface patterning techniques 10

2 Characterization and quantification of cell migration 13
 2.1 The process of cell movement 13
 2.2 Parameters to describe single cell migration 16
 2.3 Mechanical forces in cell migration 18

3 Objective 21

II Microfabricated channel structures to study cell migration   23

4 Motivation and experimental approach   25

5 Materials and methods   29
    5.1 Fabrication of micro-sized channel structures 29
    5.2 Microcontact printing of cell-adhesive lines 35
    5.3 Fabrication of nano-patterned micro-sized topographies 36
    5.4 Material characterization techniques 39
    5.5 Cell experiments 40
    5.6 Image analysis and data processing 41

6 Results   45
    6.1 Micro-fabricated migration chip 45
    6.2 Migration studies of human pancreatic epithelial cancer cells 47
    6.3 Invasiveness of medulloblastoma cells 55
    6.4 Micro-sized hydrogel channels with nano-patterned walls 58

7 Discussion   63
    7.1 Single cell migration through micro-sized channels 63
    7.2 Nanopatterned microchannels 66

III Collective cell migration on photo-switchable substrates   69

8 Motivation and experimental approach   71

9 Materials and methods   75
    9.1 Preparation of photo-switchable substrates 75
    9.2 Experimental setup 76
    9.3 Cell experiments 79
    9.4 Image analysis and data processing 81

10 Results — 85
10.1 Cluster expansion of Madin-Darby canine kidney (MDCK) cells — 85
10.2 Expansion behavior of (MCF-7) cell clusters — 100

11 Discussion — 103
11.1 Collective cell migration on photo-switchable substrates — 103

IV Conclusions and Outlook — 107

12 Cell migration through microfabricated channel structures — 109
12.1 Conclusions — 109
12.2 Outlook — 110

13 Cell migration on photo-switchable substrates — 111
13.1 Conclusions — 111
13.2 Outlook — 112

List of Figures — 113

List of Tables — 117

Bibliography — 119

A Overview of attached Videos — 135
A.1 Panc-1 cells in microchannels and on 1D lines — 135
A.2 MB cells in microchannels — 137
A.3 MDCK cells on photo-active substrates — 139
A.4 MCF-7 cells on photo-active subastrates — 142

B Additional Tables — 145

C Synthesis of photocleavable linker — 147

D MATLAB codes — 149
D.1 Binarization of phase-contrast images — 149

D.2	Angular resolved boundary positions of cell clusters	150

E List of Publications   **155**

Acknowledgement   **158**

Abstract

Cell migration is an essential characteristic of both physiological and pathological processes within the human body. In order to study the complex process of cell migration different *in vitro* model systems have been developed in the past. The challenge for all these assays is to provide the cells a substrate that mimics particular properties of the extracellular matrix (ECM) while a high control over experimental parameters and monitoring is desired. However, migration assays commonly used in cell biology and medical research are rather limited in the control over the architecture of the provided matrix on or through which the cells move or by the lack of adequate imaging devices to monitor cell dynamics.

To overcome some limitations of conventional migration assays, it was the aim of this work to develop two different methods and employ them in order to quantify migrative behavior of cells under precisely controlled *in vitro* conditions.

The first assay consists of microfabricated three dimensional (3D) scaffolds, which allow to study cell migration dynamics through confined environments via live-cell imaging. Channel structures of precisely defined dimensions were utilized to quantify the invasiveness of single cancer cells with respect to modifications of their cytoskeleton organization. In addition, dynamical migration patterns of the cells inside these confined 3D environments were analyzed and found to be significantly changed from their counterparts on flat, two dimensional (2D), surfaces.

Furthermore, it was shown that such microfabricated structures could be functionalized in the nanometer range with patterns of gold nanoparticles. Thus, the selective binding of ECM-derived ligand motifs, to the gold particles allows for mimicking specific features of the ECM in 3D.

The second assay comprises flat photo-switchable surfaces and allows for cell migration studies under precisely - spatially and temporally - controlled conditions that are dynamically adjustable during the experiments. Initially, these surfaces are repellant to cells and upon irradiation with ultraviolet (UV) light through a photomask they become locally adhesion-mediating. Migration studies with cohesive sheets of epithelial cells were performed and their expansion characteristics from geometrically confined starting conditions were quantified. As the initial size and shape of such cell sheets was varied, it was shown that only geometrical parameters like the boundary curvature of the cohesive cell sheet can directly influence and determine the collective behavior of cells.

Zusammenfassung

Die Fähigkeit von Zellen zu migrieren ist eine für den menschlichen Organismus essenzielle Eigenschaft. Migrierende Zellen übernehmen eine zentrale Rolle in sowohl lebenserhaltenden, als auch pathologischen Prozessen. Um die komplexen Vorgänge, welche während der Migration von Zellen ablaufen, zu untersuchen sind eine Reihe verschiedener *in vitro* Modellsysteme entwickelt worden. Die Hauptanforderungen an solche Untersuchungsmethoden bestehen vor allem darin, dass sowohl den Zellen einzelne Strukturelemente, welchen sie normalerweise in der Extrazellulären Matrix (ECM) ausgesetzt sind, in möglichst kontrollierter Art und Weise dargeboten werden sollen, als auch, dass die Beobachtung des dynamischen Zellverhaltens gewährleistet werden muss. Die meisten solcher Untersuchungsmethoden, wie sie vor allem in der Zellbiologie und der medizinischen Forschung angewendet werden, sind jedoch beschränkt, entweder in Bezug auf die Beschaffenheit der Oberflächen welche den Zellen zur Migration angeboten werden können, oder durch stark limitierte Möglichkeiten zur Beobachtung des dynamischen Zellverhaltens.

Um die erwähnten Einschränkungen solcher Methoden zur Untersuchung von Zellmigration zu überwinden, wurden im Laufe der vorliegenden Promotion zwei verschiedene Untersuchungsmethoden entwickelt. Diese erlauben es, das dynamische Migrationsverhalten von Zellen unter präzise definierten Bedingungen *in vitro* zu untersuchen.

Kernstück der ersten Methode bilden mikrostrukturierte Kammern welche es ermöglichen das Migrationsverhalten von Zellen durch definierte dreidimensionale (3D) Modellstrukturen in Echtzeit im Mikroskop zu verfolgen. Mit solchen definierten Kanalstrukturen wurde das invasive Verhalten einzelner Tumorzellen in Abhängigkeit von strukturellen Veränderungen des Zellskeletts bestimmt. Zudem wurde das dynamische Verhalten der Zellen während ihrer Migration durch die 3D Modellstrukturen untersucht, wobei maßgebliche Abweichungen in deren Verhalten im Vergleich zu dem auf ebenen Oberflächen beobachtet wurden.

Zudem konnte gezeigt werden, wie die Oberflächen von 3D Mikrostrukturen zusätzlich mit nanostrukturierten Mustern von Goldnanopartikeln versehen werden können. Die Goldpartikel können selektiv mit einzelnen Bindungsmotiven funktionalisiert werden, was die Möglichkeit bietet spezifische, strukturelle sowie funktionelle, Eigenschaften der ECM in 3D Modellsystemen nachzuahmen.

Die zweite entwickelte Untersuchungsmethode besteht im Kern aus photoschaltbaren Oberflächen. Mit diesen ist es möglich die Migration von Zellensembles unter - räumlich und zeitlich - präzise definierten und dynamisch anpassbaren Bedingungen zu untersuchen. Die Oberflächen wirken zunächst zellabweisend wirken, lassen sich jedoch mittels Belichtung mit ultraviolettem Licht durch eine Fotomaske in beliebigen Regionen in zelladhäsive Oberflächen umwandeln. Das Expansionsverhalten zusammenhängender Verbände von Epithelzellen wurde mit diesen schaltbaren Oberflächen untersucht. Durch systematische Veränderung der anfänglichen Größe und Geometrie der Zellverbände konnte gezeigt werden, dass diese Faktoren direkt das Maß an kooperativem Verhalten der Zellen beeinflussen und bestimmen kann.

Abbreviations

1D	one dimensional
2D	two dimensional
3D	three dimensional
BCMN	block copolymer micellar nanolithography
BSA	bovine serum albumin
CCD	charged coupled device
DAOY	human desmoplastic cerebellar medulloblastoma cell line [1]
DAPI	4',6-diamidino-2-phenylindole dilactate
DMEM	Dulbecco's modified Eagle's medium
DMSO	dimethyl sulfoxide
ECM	extracellular matrix
FBS	fetal bovine serum
EDTA	ethylenediaminetetraacetic acid
LED	light emitting diode
μCP	microcontact printing
MB	medulloblastoma
MCF-7	human breast cancer cell line [2]
MDCK	Madin-Darby canine kidney cell line [3]
MEM	minimum essential medium
miRNA	micro ribonucleic acid
MMP	matrix metalloproteinase

P1	Panc-1 cells stably transfected with keratin K8-eCFP/K18-eYFP
Panc-1	human pancreatic epithelial cancer cell line [4]
PBS	phosphate buffered saline
PEG	poly(ethylene glycol)
PDMS	poly(dimethylsiloxane)
PFA	para-fomaldehyde
PVA	poly(vinyl alcohol)
RGD	arginine-glycine-aspartic acid
RIE	reactive ion etching
RNA	ribonucleic acid
rpm	rounds per minute
r.t.	room temperature
SAM	self-assembled monolayer
s.d.	standard deviation
SEM	scanning electron microscopy
s.e.m.	standard error of means
siRNA	small interfering ribonucleic acid
SPC	sphingosylphosphorylcholine
THF	tetrahydrofuran
UV	ultraviolet

Part I

Introduction

Chapter 1

Cell migration

The term cell migration describes the biological process of cells moving themselves with respect to their environment. Such active movement of a cell is a highly complex process: The cell is orchestrating the rearrangement of the internal cytoskeleton and the cell membrane while sensing and reacting to its surrounding in order to adhere on it. Thereby, the cell builds up friction and exerts directed forces to simultaneously move from its initial position.

Cell migration plays a crucial role for a variety of pathological and non-pathological processes in every living animal: During early stages of embryogenesis collective cell movement, for example, leads to an elongation of the head-to-tail body axis [5]. In inflammation and immune-response leucocytes move rapidly from the lymph nodes through the stroma to an inflammation site where they collect information about pathogens and migrate back to the lymphatic system to present the gained information [6]. During wound healing epithelial cells are able to close a wound through coordinated migration before increased proliferation and matrix synthesis are able to completely restore the tissue [7]. In cancer metastasis, primary tumor cells invade into the surrounding tissue, enter the circulation system and migrate into healthy tissue where they can develop secondary malignant outgrowths [8].

1.1 Model systems to study cell migration *in vivo*

When it comes to study cell migration, or other cell biological questions in general, the preferred model system would be the living animal where cells interact with their native environment, typically adjacent cells and the ECM that defines the connecting tissue in animals. *In vivo* experiments, for example, with living mice are made to study cancer metastasis [9] or with the eggs of the fly drosophilia melanogaster to study early stages in development [10]. In recent years, researchers have even developed so-called *intravital* microscopy techniques which allow for monitoring cell migration in the living organism. The migration of fluorescently labeled tumor cells through native tissue, for example, can be directly studied in a living mouse [11–14]. Another model system for *in vivo* live-cell imaging is the zebrafish. This fish is transparent in the early stages of development and therefore well-suited for *in vivo* migration studies with fluorescently labeled cells [15, 16].

However, besides ethical reasons, the complexity of *in vivo* experiments leads to significant drawbacks, for example, limitations in number of experiments and poor control of testing parameters.

1.2 Model systems to study cell migration *in vitro*

To compensate for the predetermined limitations of *in vivo* experiments, *in vitro* model systems are needed to study cell migration under confined conditions.

Depending on the biological question to be studied, different aspects and parameters of cell migration are of interest, for example, migration speed, deformability of cells that squeeze through tissue, inter-cellular communication in collectively moving cell sheets, forces which the cells exert on their environment during their movement, means to sense external cues that guide directed cell migration. Migratory speeds, for example, can vary a lot between different cell types together with the underlying mechanisms: Relatively slow tumor cells migrate in vitro at speeds around 0.1–0.3 μm/min, faster fibroblasts exhibit speeds of 0.2–1 μm/min, and the fast neutrophil granulocytes are able to migrate at speeds as high as 15–20 μm/min [17]. According to this diversity in cell behavior and questions to be addressed, a variety of *in vitro* assays has been developed to study migratory behaviors of single cells and clusters of collectively interacting cells.

One of the most simple ways to characterize and quantify cell migration is to seed cells on a flat petri dish and monitor their 2D motility via live-cell imaging on a microscope. From the obtained time-lapse videos one can deduce parameters like migratory speeds and directional persistence in the movement. In experiments with cells cultured at low densities where the individual cells are separated from one another, observations of single cell migration characteristics can be made. In contrast, in experiments with cells growing at higher densities, dynamical movements within a confluent cell sheet can be studied that typically exhibit long-range interactions extending over the distance of several cell diameters.

1.2.1 Wound healing assays

A very simple and widely established method to study migration characteristics of many interconnected cells on a flat 2D surface is the so-called scratch or wound healing assay, introduced by G. J. Todaro *et al.* [18] that mimics the closure of a wound. To do so, adherent cells are cultured in a petri dish until confluency is reached. With a sharp razor blade or the tip of a pipette cells are mechanically scratched away, leaving a thin corridor, several tenths of micrometers thick, in the confluent cell sheet, which represents the wound as depicted in Figure 1.1. The closure of this wound can be followed under the microscope and the speed with which the remaining cells re-occupy the free area can be quantified [19]. Better control over the shape of the wounded area can be gained with large arrays of scratching devices [20] or generating circular wounds [21].

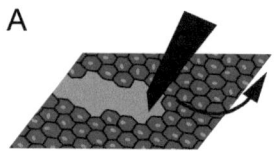 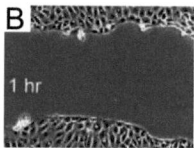 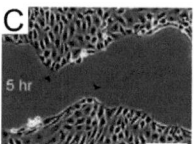

Figure 1.1: Illustration of the wound healing assay. (A) A device like a razor blade or pipette tip is used to mechanically disrupt a confluent layer of cells and generate an artificial wound. (B-C) Example of an artificial wound in a scratch assay. Phase-contrast images were taken 1 hours (B) and 5 hours (C) after wounding a confluent layer of epithelial cells. After 5 hours it is clearly visible how the boundaries of the separated cell sheets move towards each other to close the wound. Images taken from [22]; scale bar, 200 μm

1.2.2 Cell invasion assays

Both, cancer cells which are involved in tumor metastasis, as well as leukocytes which are responsible for an immune response, are able to move independently through tissue in a directed way. Many attempts were made in the past to study such directed and invasive migratory behavior.

One of the most common and also oldest experiments commercially available is the so-called Boyden chamber, also known as transwell migration assay [23]. The Boyden chamber consists of two reservoirs separated by a thin porous membrane with defined pore diameters usually smaller than the cell diameters [24] as depicted in Figure 1.2. The upper reservoir is filled with the cell suspension and after a certain incubation time the amount of cells which migrated through the micropores to the lower side of the membrane is counted. The setup is fairly simple to prepare and also available for 96-well plates which allows for a relatively high throughput and statistical evidence of the experiments. Additionally, the porous membrane can be coated with a layer of a collagen-based gel mimicking the ECM. The lower reservoir can be also filled with a chemoattractant to study the influence on directed migration of the cells through the pores towards the test substance [25].

One of the main drawbacks of the Boyden chamber, however, is that it lacks

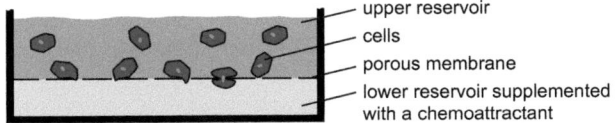

Figure 1.2: Illustration of the Boyden chamber invasion assay. A porous membrane is separating two reservoirs that are filled with medium and optionally supplemented with a chemoattractant. Cells seeded in the upper reservoir adhere to the membrane and migrate eventually through the pores. At the end of the experiment the membrane is removed and the cells that migrated through the pores are counted.

information about the real migration speed of the cells and the dynamics how the cells actually squeeze through the pores.

1.3 Surface functionalization to mimic the extracellular matrix

The extracellular matrix (ECM) consists of many different fibrous proteins and polysaccarids like collagen or fibronectin and proteoglycans, to name only a few. These components not only provide a mechanical scaffold to the tissue, but also expose a wide range of ligands that are recognized by receptors on the cell membrane. The interactions between cell membrane receptors and ligands presented by the ECM are essential to maintain functionality of single cells and the whole tissue [26, 27]. Therefore, it is of importance to control the surface properties in *in vitro* cell experiments by either binding whole ECM-derived filamentous proteins, or small molecules that represent single ligand motifs to the surface. One prominent representative of such adhesion-mediating ligands is the tri-peptide arginine-glycine-aspartic acid (RGD) which binds, especially in its cyclic form, selectively to the membrane receptors $\alpha_v\beta_3$-integrin and $\alpha_v\beta_5$-integin [28].

In addition to surface bio-functionalization methods that try to mimic the ECM, coatings have been developed that effectively inhibit adsorption of proteins and prevent cell adhesion. Typically, highly hydrated polymers like poly(ethylene glycol) (PEG) [29–31], poly(vinyl alcohol) (PVA) [32], Pluronic, which is a triblock-copolymer of PEG-poly(propylene glycol)-PEG [33,34], or highly hydrophobic and non-polarizable polymers like poly(tetrafluorethen) (Teflon) [35] are used to provide biologically inert surfaces.

The requisites and complexity of ECM-mimmicking surface functionalizations is highly dependent on the studied cell type, the composition of the surface and the biological question that shall be tackled. The field of surface bio-functionalization is wide [36] and only a short glimpse of different coating methods and functionalization techniques can be given here, with examples illustrated in Figure 1.3.

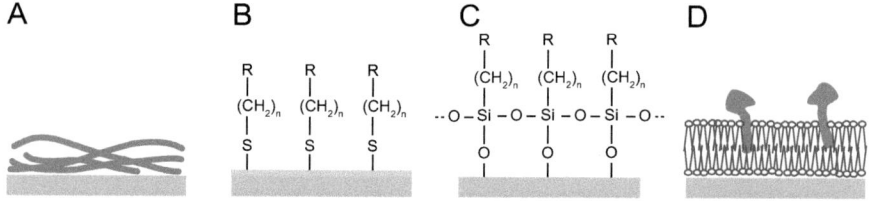

Figure 1.3: Illustration of different surface functionalization techniques. (A) Physisorbtion of long protein filaments to the glass surface that adhere via electrostatic interactions. (B) Selective binding of thiols to a thin layer of gold that was sputtered on a glass surface. Once attached to the surface the single molecules are still able to move their positions and form highly ordered self-assembled monolayers (SAMs). Typically, the rest of the molecule is already functionalized or it contains reactive groups that can be bio-functionalized. (C) Direct binding of organosilane molecules to a glass surface where they form a inter-connected monolayer. Like in (C), the terminated groups are either functionalized or can be functionalized in a following step. (D) Lipid bilayer with incorporated membrane binding proteins which are able to diffuse on the surface. Illustrations are not to scale.

Physisorption The most simple method to bio-functionalize a non-biologic surface is to incubate it with a solution of ECM derived fibrillar proteins like fibronectin or collagen and let the proteins adsorb to the surface via electrostatic interactions (physisorption). Under normal cell culture conditions animal derived serum like fetal bovine serum (FBS) is added to the cell medium which, among other components, already contains high amounts of ECM derived proteins. For most culture conditions and experiments with cells on normal glass slides or plastic dishes, physisorption of the fibrous proteins in the serum is sufficient to let the cells adhere and spread on the surface. However, in a physisorption process, little control of the orientation of the adhering compounds with respect to the surface is possible, compared with the following bioconjugation techniques.

Thiol coupling agents on metal surfaces Solutions of thiol-containing molecules can be brought in contact with metal surfaces, like gold films, in order to cover large surface areas with self-assembled monolayers (SAMs) [37–39]. The nature or the metal-sulfur bond is not completely understood and still controversially discussed in literature. In contrast to a typical covalent bond, the thiol molecules are, for example, still able to laterally diffuse on the surface which allows them to form the highly ordered SAMs [39]. Also disulfides, as found in many cystein containing proteins, are able to link to a gold surface without prior reduction to thiols. Gold films of only a few nanometer thickness can be easily obtained via physical vapor deposition of gold atoms on glass slides, for example.

Silane coupling agents Silane chemistry allows for directly coating glass surfaces with monolayers of functional linking-molecules. Silanes that contain at least one bonded carbon atom are called organosilanes. They can have hydrogen, oxygen, or halogen atoms directly attached to the silicon atom core. Some of these derivates are highly reactive and can be used to form covalent linkages with other molecules or surfaces upon hydrolysis. Typically, the organic part bound to the silicon atom terminates with a functional group that allows conjugation of the organosilane to other organic compounds while the silane reactive groups are chosen to only couple to inorganic substrates like glass surfaces [36].

Lipid bilayers Another way to mimic ligand distributions of the ECM or adjacent cell membranes is to use lipid bilayer-based surface coatings or vesicles. A rather simple and straight-forward method to generate lipid bilayers on a flat surface with a Langmuir-Blodgett balance where a substrate is first immersed into and then pulled out from a aqueous solution with a compressed monolayer of lipid molecules at the air/water interface [40]. Membrane-binding ligands can be incorporated into the lipid bilayer where they are able to diffuse on the surface [41–43].

1.4 Surface patterning techniques

In order to mimic ECM cues for the cells, it is not only important to provide specific ligands on the surface, but also to control their local distribution in a controlled way.

Microcontact printing Microcontact printing (μCP) is a well-established method commonly used to pattern flat surfaces with adhesive and non-adhesive areas at a resolution down to a few micrometers [44–48]. Therefore, a thin layer of molecules or polymers is transferred via a silicon rubber-based stamp to a substrate on which it remains either by physisorption or covalently binding. For cell migration studies, cell-adhesion mediating proteins or protein fragments are usually printed in the desired pattern on a surface and the non-printed areas are passivated after the printing process [49].

Other micropatterning techniques Completely passivated surfaces can be also used for patterning with protein or cell adhesive areas. In this case, the passivating organic molecules like PEG or PVA are either decomposed by irradiation with UV light [50] in selected regions or through focusing a high energy laser onto them [32].

Block copolymer micellar nanolithography Even the control and variation of single ligand-to-ligand distances on a nanometer scale is possible. With the help of the so-called block copolymer micellar nanolithography (BCMN) [51–53] it is possible to generate patterns of hexagonally ordered nano-sized particles on flat silicon wafers or glass slides with an area of up to several square centimeters. In this method, the substrate, typically a glass slide or a silicon wafer, is immersed in a micellar solution of a diblock-copolymer. The poly(styrene)-block-poly(2-vinylpiridine) copolymer is dissolved in an organic solvent like toluene where it forms a solution of inverted micelles, which can be loaded with a metal salt, e.g. tetrachloroauric acid ($HAuCl_4$). When the substrate is pulled out of the micellar solution, the micelles arrange themselves in a highly ordered monolayer. Finally, reduction of the gold salt to elemental gold while simultaneously removing the organic polymer chains in a plasma oven generates patterns of hexagonally ordered gold particles with a diameter of 5–10 nm and an inter-particle spacing adjustable between roughly 20–200 nm. The chemical contrast between the gold particles and the silicon oxide on the rest of the surface can then be used to selectively bind functional molecules to the bi-functioal surface: Organosilanes can selectively bind to the glass surface, while thiols only bind to the surface or the gold particles. In recent years, protocols were developed that allow to transfer the hexagonally ordered gold particles to the surface of a PEG-based hydrogel which is adjustable in its stiffness [53]. It was even shown that such nano-patterned hydrogels can be reversibly stretched, which allows to dynamically change and control the inter-particle distance of the gold particles on a nanometer scale [54].

1.4.1 Surface topography and 3D environments

Topography plays an important role for cellular migration behavior. Cell migration on topographically structured surfaces has been extensively studied, for example, on micro-sized and nano-sized grooves [55, 56], on top of micropillars [57], and through arrays of micropillars [58]. With increasing depth of the topography the model systems resemble more and more the complex 3D architecture of the ECM. Examples of *in vitro* studies of cell migration in 3D matrices are cells embedded in collagen gels [59–61] or synthetic PEG hydrogels [62, 63] that can be locally degraded by cancer cells expressing matrix metalloproteinases (MMPs).

Migration mechanisms have been found to be varied for cells exposed to either a flat 2D surface, a 3D gel matrix or highly restricted one dimensional (1D) adhesive lines, as schematically depicted in Figure 1.4: Migration characteristics of fibroblasts, such as the speed and the motion pattern, were found to be more

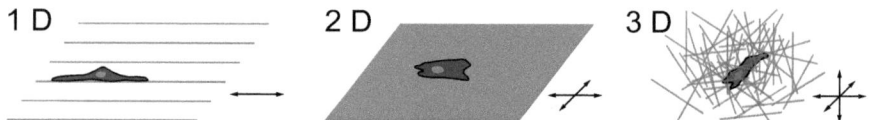

Figure 1.4: Schematic of cell migration in different dimensions. (1D) Cell migration on thin adhesion-mediating lines can be considered as one dimensional, since the directionality in movement is restricted to only two directions. (2D) Cells on a flat surface like a conventional petri dish are able to move in all directions on the surface. (3D) Cells inside a 3D matrix, a gel for example, can move in all three dimesions. Arrows indicate the possible directions in movement.

similar in 1D and 3D than in 2D environments [32]. It was also reported that leukocytes can switch their migration mechanism from an adhesion-dependent migration mode on a flat surface to an adhesion-independent mechanism when seeded in a 3D environment [61].

Chapter 2

Characterization and quantification of cell migration

2.1 The process of cell movement

Like a car needs a motor and tires to move on a street, also cells need mechanical elements in order to move through the ECM. Mechanical stability and structure in the case of a cell is provided via the cytoskeleton, which consists of three different kinds of polymers: Actin, intermediate filaments and microtubuli. Filamentous actin forms long and stiff fibrils and, together with myosin motor proteins, it assembles to contractile structures (*the motor of a car*). Intermediate filaments play an important role in defining the mechanical stiffness and elasticity of the whole cell and the nuclear envelope [64–66], while microtubuli are important to maintain the cell structure, its polarization, serve as a platform for intracellular transport via self propelling motor proteins moving along them, and play an important role during mitosis [67, 68]. The connection to the extracellular space is mediated via specific cell membrane receptors like integrins which are able to attach to the ECM to build up friction and transmit forces (*the tires of a car*). Finally, the ECM is exposing ligands to the membrane receptors, serves as a mechanical scaffold and guides cell migration (*the street for a car*).

The ability of cells to actively move and delocalize their cell body is a highly orchestrated process. The mechanisms, which cells use to move themselves with respect to their environment, can be classified into the two categories of amoeboid and mesenchymal migration [69].

Amoeboid migration characterized by short-lived and relatively weak interactions with the substrate. It is mostly used by nonresident cells like leukocytes and stem cells who are able to enter and rapidly move through many organs [69].

Mesenchymal migration, in contrast, involves specific cell-ECM interactions and a sequence of independent steps which, subsequently carried out, allow for cells to move themselves with respect to the ECM. This migration pattern, which is also known as the *push-and-pull* mechanism has been first described for fibroblasts migrating on a flat surface [70], but it is also observed in the movement of many other cell types in 2D as well as in 3D surroundings [71]. The following four steps, as depicted in Figure 2.1, are characteristic for mesenchymal migration: (1) Initial cell polarization is driven by localized actin polymerization which promotes membrane protrusions at the leading edge, so called pseudopods (*greek: false feet*), which, depending on their shape can be separated into the broad lamellimpodia or the thin filopodia. (2) When the protruding pseudopods get in contact with ligands of the ECM, specific adhesion receptors, e.g. integrins, attach to the ligands, forming focal contacts at the leading edge. Depending on the cell type and the constitution of the ECM, MMPs can be activated in this step in order to locally degrade adjacent ECM proteins via proteolysis to widen the available space, especially for the migration through a 3D scaffold. (3) During, or shortly after the ligand-integrin binding, actin filaments are stabilized and anchored to the membrane via cross-linking proteins and contractile proteins like myosin II. Shortening of the actin fibers via myosin driven tightening leads to a local cell contraction and a movement of the cell body. (4) Finally, the detachment of the cell membrane from the ECM at the trailing edge completes the translocation of the whole cell, before the whole process is repeated.

However, it should be kept in mind, that these four steps do not represent a stereotypic program but rather provide an adaptive platform that undergoes modifications, dependent on the cell-type and the composition of the ECM [72].

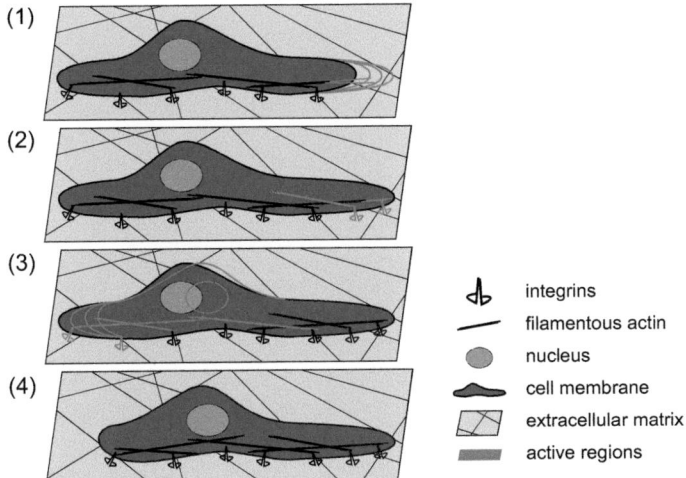

Figure 2.1: Single steps of cell migration. (1) Outward directed actin polymerization leads to membrane protrusions. (2) Binding of the membrane receptors to the ECM and formation of focal adhesions at the leading edge. (3) Local actomyosin contraction in the rear of the cell leads to movement of the cell body. (4) Detachment from the ECM at the rear of the cell leads to a translocation of the cell. Active regions in each step are highlighted in red.

2.2 Parameters to describe single cell migration

Single cells that actively migrate on a surface without any gradient presented on the surface or in the medium show a random movement. In order to quantitatively describe such movement, different parameters can be taken into account, with some of the most important ones described below.

Velocity (v) The velocity of a migrating cell is obtained from the accumulated distance (d_{accum}), which represents the whole length of a cell track, and relating it to the duration of the cell track, the time interval Δt, as illustrated in Figure 2.2.

$$v = \frac{d_{accum}}{\Delta t} = \frac{\sum_{k=1}^{i} r_{k+1} - r_k}{t_{k+i} - t_k} \qquad (2.1)$$

However, if the movement is not linear, the velocity is not sufficient for describing the cell movement, as a high velocity would also result from a cell that is moving rapidly back and forth within a small area without being significantly delocalized after a longer period of time.

Figure 2.2: Cell track from a random walk movement. Real cell track (green line) with the positions indicated where the cell position was measured (green dots). (A) Accumulated distance of the cell track is indicated by the black line. (B) Euclidian distance of the cell track is indicated by the black line. Figure adapted from [73].

Directionality (D) Directionality in cell movement is calculated by comparing the euclidian distance (d_{euclid}), the shortest distance between beginning and end of a cell track, with the accumulated distance (d_{accum}). It is a measure of directness of a cell trajectory. A directionality of $D = 1$ would refer to a migration path along a straight line.

$$D = \frac{d_{euclid}}{d_{accum}} = \frac{r_{k+i} - r_k}{\sum_{k=1}^{i} r_{k+1} - r_k} \quad (2.2)$$

Directional persistence time (P) and root-mean-square speed (S) A common feature found for many cell tracks is that cell movement persists in the same direction over short times, but over longer times directional changes occur in a random way. This observation has led several investigators to mathematically model random cell movement as a correlated random walk [73, 74]. Although some of the underlying assumptions of these models vary slightly, a common feature of each is that the cell velocity, v, at time t, has a correlation which decays exponentially with time. Thus, the velocity autocorrelation function, $G_v(\tau) \equiv \langle \mathbf{v}(t+\tau) \cdot \mathbf{v}(t) \rangle$, can be written as follows [73]:

$$\langle d^2(t) \rangle = 2S^2 P[t - P(1 - e^{-1/P})] \quad (2.3)$$

The cell track is assumed to consist of a sequence of n cell positions associated with a series of increasing time points differing by a constant time increment, Δt. If $\mathbf{r}_k$ represents the position vector at the kth time point, as depicted in Figure 2.3, then the cell displacement over the time interval $t_i \equiv i\Delta t$, from $\mathbf{r}_k$ to $\mathbf{r}_{k+i}$, is:

Figure 2.3: Cell track with measured time points. (A) Real cell track (green line) with the positions indicated where the cell position was measured (green dots) (B) Obtaining displacement data ($i = 3$) from overlapping intervals. The dashed lines represent samples of displacements used in the averaging process. Figure adapted from [73].

$$\mathbf{d}_{ik} \equiv \mathbf{r}_{k+i} - \mathbf{r}_k \quad (2.4)$$

Let x_{ik} be the squared displacement from $\mathbf{r}_i$, to $\mathbf{r}_{k+i}$:

$$x_{ik} \equiv \mathbf{d}_{ik} \cdot \mathbf{d}_{ik} \qquad (2.5)$$

Then, x_{ik} is considered a random variable with expected value $\eta_i \equiv \langle x_{ik} \rangle = \langle d^2(t_i) \rangle$, where η_i is the theoretical mean-squared displacement over t_i given by Equation 2.3. Now, there are different possibilities to calculate the measured mean-squared displacement $\bar{x}_i$ at the corresponding time interval t_i from time-lapse video data. The most simple way is to average several squared displacements over the cell track as illustrated in Figure 2.3. To maximize the total number of samples from a single track, n_i, one can average squared displacements from overlapping time intervals ($x_{i,k+1}$, $x_{i,k+2}$, $x_{i,k+3}$, etc.):

$$\bar{x}_i = \frac{1}{n_i} \sum_{k=1}^{n_i} x_{ik} \qquad (2.6)$$

with $n_i = (n - i)$. Plotting the obtained mean square displacements versus the time, typically show an exponential increase in $\langle d^2(t) \rangle$ and fitting the data to Equation 2.3 one can obtain the parameters of directional persistence time (P) and root-mean-square speed (S) which give characteristic information about the cell migration behavior [73].

2.3 Mechanical forces in cell migration

Cell migration, either of single cells or cell clusters, is dependent on exerting a traction force on the environment via acto-myosin-based contractions of the cytoskeleton and cellular adhesion sites in order to actively delocalize. Experimental methods have been developed to measure such forces on a micro-scale resolution [75]. A flat surface, for example, can be coated with a layer of a soft gel with a stiffness similar to the one found in real tissue (~0.1-10 kPa) [76, 77] with fluorescently labeled micro beads. Once adherent cells attach to such a flexible surface and start to migrate, the changes in bead localization are monitored from which the forces, that the cell exerts on the surface, can be deduced [78–81]. Typical forces that cells exert on a surface while they are migrating can differ a lot among different cell types and they are in the following range: 0.1-85 nN/μm^2 [82, 83]. Force measurements of cells embedded in a 3D matrix is much more complex. To this point most reports describe only morphological changes of the cell surrounding matrix [84] and methods to directly deduce mechanical information from cells, cultured inside a gel, for example with Young's moduli of ∼0.6 and ∼1 kPa, are still rather sophisticated [85]. Additionally, migration mechanisms, especially traction force generation, of cells moving through a 3D matrix might be fundamentally different to the force generation on a flat surface. Theoretical models, for example, propose a mechanism of cell motility in confinements, that only relies on the coupling of actin polymerization at the cell membrane to the geometric confinement and does not require any specific adhesion to ECM ligands [86].

Chapter 3

Objective

Cell migration is a complex and highly orchestrated process which is, on one side, controlled by intracellular processes and, on the other side, highly dependent on the mechanical and chemical composition of the ECM or the presence of neighboring cells. Hence, *in vitro* model systems, too, are required to mimic characteristic attributes of the ECM, depending on which biological questions are to be addressed during the experiment.

It was the objective of this work to develop specific *in vitro* model systems that provide particular attributes for migration studies and use them to characterize and quantify cell migration under highly controlled conditions. Two distinct *in vitro* model systems were developed that should enable the investigation of the following biological questions:

- How does the organization of the keratin cytoskeleton influence the deformability and invasiveness of single pancreas tumor cells? How are migration dynamics changed when cells are either moving inside confined 3D environments or on flat 2D surfaces?

- How are migration dynamics of a collective epithelial cell sheet influenced by its initial size, shape? Does a critical size of the cell sheet exist, at which intrinsic cohesive properties start to dominate over single-cell behavior; that is to say, when does an ensemble of cells start to show collective properties?

In the following parts II and III of this thesis, two different model systems are described that were developed to study either the invasiveness of single cells in a confined 3D environment or to study collective migration properties of cohesive cell clusters. Each part starts with a biological motivation and a description of the experimental approach, followed by the materials and methods section, the results section including the cell experiments, and the discussions. In the last part of the thesis, the conclusions are presented together with an outlook.

Part II

Microfabricated channel structures to study cell migration

Chapter
4

Motivation and experimental approach

Migration of tumor cells has been extensively studied due to its importance in the process of cancer metastasis [8, 87]. Tumor cells in a primary tumor are able to abandon their initial environment and migrate through the surrounding parenchyma, enter the circulatory system and invade other, healthy tissues. On this journey, the cells need to regulate their migratory and invasive behavior and are exposed to a variety of mechanical interactions like shear stress and deformation [75,88,89]. Remarkably, it was reported that upon pharmacological inhibition of MMPs, cells, instead of proteolytically degrading their local environment, were forced to "squeeze" through a collagen fiber network accompanied by drastic cell and nuclear deformation upon migration [59]. These deformations require substantial reorganization of the cytoskeleton and other organelles. In fact, the mechanical properties of a cell are likely to be crucial for its migratory behavior in a given environment [88].

For example, migration mechanics of human pancreatic adenocarcinoma cells Panc-1, a cellular model system expressing mainly keratin 8 and 18 as intermediate filaments, are highly dependent on the cell's keratin cytoskeleton organization. Addition of the bioactive phospholipid sphingosylphosphorylcholine (SPC) leads to a substantial modification of the keratin network towards a perinuclear rearrangement [90, 91]. This results in reduced cell stiffness and enhances the ability of cells to squeeze through porous membranes in a Boyden chamber assay. Thus, it is suggested that SPC enhances the metastatic potential of pancreatic tumor cells [92].

One objective of this thesis was to develop a new experimental setup that would allow for the investigation of such invasive cell behavior in more detail, getting over the limitations given by the Boyden chamber assay, that does not provide any information about migration dynamics. Therefore, the new setup had to meet with the following prerequisites:

- Providing a well-defined 3D scaffold/architecture where the cells are able to migrate and squeeze through,

- Feasibility to conduct live-cell imaging with a statistically relevant amount of cells and experiments,

- Possibility to chemically modify the surface properties in order to mimic special ECM functionalities,

- Suitability for long-term observations over several hours or days and

- Assuring high reproducibility for independent experiments.

Thus, a migration assay was developed whose key features are micro-sized channel structures of various widths, lengths, and heights that are coated with ECM proteins and can be accessed by the cells. The channel structures are cast via a replica molding process using the transparent polymer poly(dimethylsiloxane) (PDMS) and are bound to a glass slide. As arrays of neighboring microchannels are arranged in close proximity, cell migration dynamics at multiple channels can be simultaneously monitored using an automated microscope.

To compare directed migration in a confined 3D channel with directed migration on thin quasi-1D lines, where only a small part of the cell membrane is in contact to the surface, a microcontact printing (μCP) process was applied where the flat surface of a passivated petri dish was patterned with adhesion mediating fibronectin lines.

The PDMS-based channel structures can be coated via physisorption of protein filaments like collagen or fibronectin that exhibit specific ligands to the cells and mimic the ECM. The physisorption of the proteins generates homogeneously coated surfaces which is sufficient for a big variety of *in vitro* applications.

To further increase the control over concentration and spacial distribution of ECM-mimmicking ligands on the microstructured surface, a technique was developed that combines the replica molding of micrometer-sized features of a PEG-based hydrogel with the transfer of nano-sized gold particle patterns on the surface obtained via BCMN. Such 3D microstructured and protein repellant hydrogels, decorated with different patterns of gold particles on the surface, can be selectively functionalized with ligands derived from the ECM or other molecules that bind to the gold particles.

Chapter
5

Materials and methods

5.1 Fabrication of micro-sized channel structures

A two-step photolithography process with high aspect ratio features of the mechanically robust SU-8 photoresist was applied. Casting molds with structures of two different heights were used for replica molding with the transparent PDMS polymer. The whole assembly process of the migration chips is summarized in Figure 5.1 and the single steps are described in the following paragraphs.

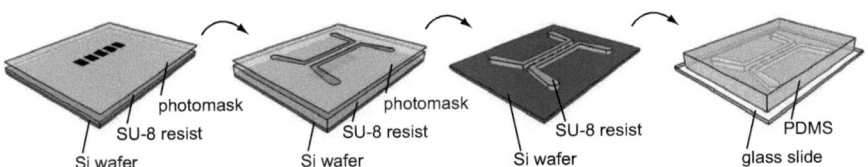

Figure 5.1: Overview of the two-step photolithography and replica molding. First, a thin layer of photoresist is exposed with a photomask that determines the width and height of the channels. In the second exposure step a thicker layer of photoresist is applied to control the channel lengths and the height of the reservoirs. The wafer with the developed photoresist is used for replica molding with the transparent PDMS polymer. The polymerized replica is peeled off the casting mold and covalently bound to a glass slide.

5.1.1 Photolithography

Migration chips with two different layouts were developed. In one of them, 40 channels and the reservoirs are incorporated into a closed microfluidic framework, while in the other 320 channels of different size are located in an open configuration, that can be directly filled with the cells and medium using a pipette. Figure 5.2 shows an illustration of the chrome masks that were used in the two-step lithography process for each migration chip. The layouts of the masks were drawn with AutoCAD (Autodesk, USA) and the tailor-made chrome masks were purchased from Masken Lithographie & Consulting (Germany).

The first, thin layer of SU-8 10 photoresist (MicroChem, USA) was spin-coated onto the freshly cleaned silicon wafer and defined the height of the microchannels (11 μm); detailed parameters of the photolithography steps are summarized in Table 5.1. After soft baking, the channel widths were defined (7–15 μm) via exposure with the first chrome mask on a mask aligner (MJB4; SUSS, Germany) using the G-band of a mercury arc lamp (HBO 350 W; Osram, Germany). Subsequently after the post exposure bake, a second, much thicker, layer (150 μm) of SU-8 2075 photoresist (MicroChem, USA) was spin-coated on the first layer. Exposure with the second mask, that was aligned to the structures from the first exposure step, allowed for the building of two reservoirs for media on both sides of the channels and thereby also determined the channel length (50, 150 and 300 μm). The reservoirs were relatively high (150 μm) in order to provide the cells with enough media for more than 24 hours. After the second post-exposure bake the structured wafer was developed (mr-Dev 600; Microresist, Germany) and hard baked. All solvents used in the lithographic process were purchased in Selectipur® grade (BASF, Germany).

5.1.2 Replica molding and chip assembly

The surface of the structured master wafer was coated with a monolayer of a perfluorated alkylsilane to facilitate the following lift-off process of the PDMS. For that, the surface of the freshly structured silicon wafer was activated under a reactive

FABRICATION OF MICRO-SIZED CHANNEL STRUCTURES   31

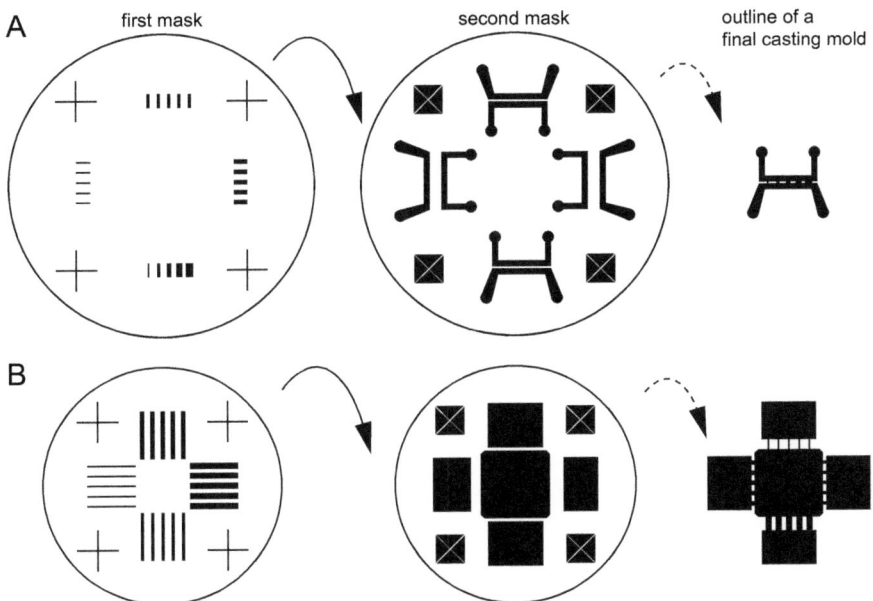

Figure 5.2: Illustaration of tailor-made chrome masks. The first mask is confining the channel widths while the second defines the channel lengths and the shape of the reservoirs. Alignment masks in the corners are needet to align the second photomask correctly on the wafer. The schematic is not to scale and black areas indicate the transparent regions of the chrome mask. (A) Layout of the two photomasks that were used to obtain the casting mold for the migration chip that is incorporated into a closed microfluidic system. Four migration chips are simultaneously processed on one three inch wafer. In each of the four migration chips, 40 microchannels are contained. By rotation of the first mask, with respect to the second one, the amount of possible channel dimensions can be increased. (B) Layout of the two photomasks used for the migration chip that can be directly filled with a pipette. One migration chip at a time is processed on a a two inch wafer with a total of 320 channels.

Table 5.1: Photolithography parameters for SU-8 resist. Process steps to obtain thicknesses of 11 and 150 µm

SU-8 10 photoresist		
clean I	3 min	acetone
clean II	3 min	isopropanol
rinse		deinoized water
dry I		blow dry with nitrogen
dry II	10 min	200°C
spread cycle	5 sec	500 rpm; acceleration 100 rpm/sec^2
spin cycle	35 sec	3000 rpm; acceleration 300 rpm/sec^2
pre-bake	2 min	65°C
soft bake	5 min	95°C
exposure	5 sec	first mask, hard contact
post exposure bake 1	2 min	65°C
post exposure bake 2	4 min	95°C
SU-8 2075 photoresist		
spread cycle	30 sec	500 rpm; acceleration 100 rpm/sec^2
spin cycle	60 sec	1000 rpm; acceleration 200 rpm/sec^2
soft bake	6 h	95°C (ramp from 65°C); slow cooling
exposure	10 sec	second mask, hard contact
post exposure bake	15 min	95°C (ramp from 65°C); slow cooling
develop	4 min	with intermediate sonication steps
rinse		isopropanol
dry		blow dry with nitrogen
hard bake	10 min	200°C and slow cooling to r.t.

oxygen plasma (5 sec with 150 W and 0.1 mbar O_2 in a PVA TePla100 plasma oven; TePla, Germany) and subsequently placed inside an exsiccator together with two drops of 1H,1H,2H,2H-perfluorooctyltrichlorsilane (#448931, Aldrich, Germany) placed on a glass slide next to the wafer. In order to saturate the atmosphere with the silane, a vacuum was applied until the boiling pressure of the perfluoroalkyltrichlorosliane was reached ($\sim 10^{-3}$ mbar) and left to incubate for at least 2 hours.

For replica molding, the PDMS prepolymer (Sylgard 184; Dow Corning, Germany) consisting of a base and a curing agent, was mixed in a ratio of 10:1. In order to get rid of bubbles, the mixture was degassed in an exsiccator under vacuum until no more bubbles appeared. This freshly prepared mixture was cast over the structured silicon wafer placed in a petri dish, evacuated again until the PDMS was completely degassed and finally cured overnight at 65°C in a convection oven. The cured replica was cut out with a scalpel, gently peeled off and examined under a light microscope for the quality of the microstructures.

To seal the PDMS-based replica onto a glass slide, glass slides were cleaned in an Extran MA01 (Merck, Germany)/Water (1:3) bath, sonicated for five minutes, subsequently rinsed with deionized water, dry-blown with nitrogen, and heated for 30 min at 110°C in order to remove the water on the surface. The glass and PDMS surfaces were activated for 20 seconds under an oxygen plasma with a pressure of 0.1 mbar and a power of 150 W. Gently pressing the PDMS on the glass and heating to 75°C for 10 min allowed for a permanent and irreversible bonding of the chip. Finally, the chip was again examined under a light microscope to confirm the proper bonding of the PDMS microchannel structures onto the glass.

5.1.3 Preparation of the migration chip for cell experiments

For the migration chip connected to a microfluidic system, blunt syringe needles (30-G; Transcoject, Germany) were pushed through the PDMS down to the reservoirs to connect them to microfluidic tubings and a 1 mL syringe. The needles were sealed with epoxy glue (Uhu plus schnellfest; Uhu, Germany) in order to prevent leakage. Prior to use, each chip was heated for 1 hour to 110°C for sterilization, carefully flushed with water to remove all air bubbles, followed by phosphate buffered saline (PBS), before a 50 µg/mL bovine or human fibronectin solution (#33010018, #33016015; Life Technologies, Germany) was injected. In migration chips where the channels were not incorporated into a microfluidic circuit, the channels were filled with fibronection solution directly after bonding to the glass while the activated PDMS was still hydrophilic. The fibronectin solution was incubated for 3 hours at room temperature (r.t.) or over night at 4°C, permitting physisorption of the protein filaments on the surfaces and making the PDMS cell adhesive. Directly before the experiments started and cells were introduced into the migration chip, the fibronectin solution was exchanged by normal cell culture medium and the whole migration chip was pre-warmed at 37°C.

5.2 Microcontact printing of cell-adhesive lines

For the study of cell migration on line structures, ultra-low attachment culture dishes (Corning, USA) which are covered with a thin hydrogel layer, were patterned with cell adhesion-mediating fibronectin lines via μCP. In the μCP process a PDMS replica was used as a stamp that molded from a photolithographically structured wafer. The photolithography parameters were the same ones as described for the first step in section 5.1.1 and the linear grooves on the stamp were 7 μm wide, 11 μm high, and 5 mm long. A 50 mg/mL fibronectin solution (#33016015; Life Technologies, Germany) was used for the printing with 25% of the fibronectin fluorescently labeled with Atto488 (Atto-Tec, Germany) in order to visualize and control the printing efficiency. The single steps of the μCP process are summarized in Table 5.2. Subsequently after printing the fibronectin lines, dishes were filled with culture medium and ready for cell experiments.

Table 5.2: **Protocol for microcontact printing.** This process allows for patterning the surface of ultra-low attachment dishes with ECM proteins.

step	duration	
1		take a freshly cast stamp
2	30 sec	immerse in acetone
3	60 sec	immerse in isopropanol
4		blow dry with argon
5	60 sec	incubation with fibronectin solution
6	5 sec	immerse in PBS
7		blow dry with argon
8 + 9		print twice
10		repeat from step 2

5.3 Fabrication of nano-patterned micro-sized topographies

In order to obtain micro-sized grooves with the walls decorated with patterns of gold nanoparticles, the following approach was developed. Microstructures were first anisotropically etched into silicon wafers with reactive ion etching (RIE). Then, the etched microstructures were patterned with gold nanoparticles via block copolymer micellar nanolithography (BCMN). In a following step, the gold particles on the surface of the microstructured silicon wafer can be transferred and covalently bound to a biologically inert PEG hydrogel as depicted in Figure 5.3.

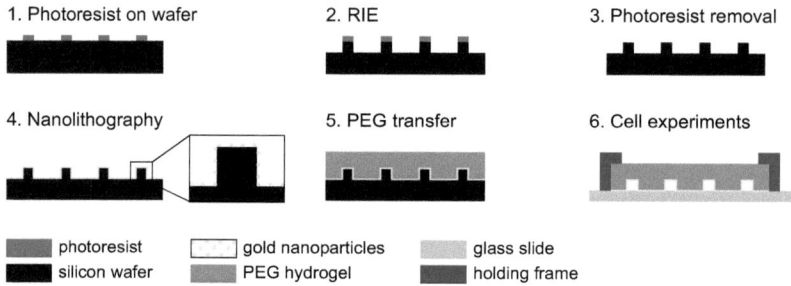

Figure 5.3: Experimental approach to fabricate nano-patterned microstructures. (1.-3.) First, microstructures with vertical side-walls are anisotropically etched into a silicon wafer using a RIE process and a photoresist that is locally protecting regions of the silicon wafer that shall not be etched. (4.) Then, the remaining microstructured surface is covered with patterns of gold nanoparticles. (5. and 6.) In a replica molding process the microstructures can be transferred together with the gold particles to a PEG-based hydrogel and used for cell experiments.

5.3.1 Silicon microstructures

Silicon wafers (two inch, ⟨100⟩ orientation, p-type; Siegert Consulting, Germany) were patterned with a 1.5-1.8 μm thin layer of the photo-reversal photoresist AZ 5214E (Clariant, Germany) with the single processing parameters summarized in Table 5.3. The AZ 5214E is called photo-reversal because it can be used as a positive or negative photoresist, depending on the processing steps; in these experiments it was used as a positive resist. For the developing step, a solution of AZ 351B (Clariant, Germany) with deionized water (1:4) was used.

Table 5.3: Photolithography parameters for AZ 5214E resist. This protocol leads to resist thicknesses of 1.5-1.8 μm.

AZ 5214E photoresist		
clean		blow with nitrogen
spin cycle	20 sec	4000 rpm; acceleration 570 rpm/sec^2
soft bake	2 min	100°C
exposure	1.5 sec	hard contact
wait	5 min	r.t.
reversal bake	1 min 45 sec	120°C
photo-reversal	10 sec	flood exposure
develop	2x15 sec	developer solution
rinse		H$_2$O
dry		blow dry with nitrogen

The silicon wafers, pre-structured with photoresist, were directly placed into a dry reactive ion etcher (Plasmalab 80 Plus; Oxford Instruments, UK) for anisotropic etching (RIE). In order to obtain rectangular walls of the etched features, multiple repetitions of etching and passivation steps were applied. For silicon, fluorine is the primary etching agent, supplied by the SF$_6$ gas, while CF$_2$ supplied by the CHF$_3$ gas is passivating the sidewalls [93, 94]. The parameters of etching and passivation steps are summarized in Table 5.4. The wafers were processed at -10°C and 100 iterations of the etching cycle resulted in groove depths with 10 μm deep vertical sidewalls.

After the RIE process, the remaining photoresist was dissolved in dimethyl sulfoxide (DMSO) and the structures were thoroughly rinsed in deionized water. Organic residues were etched away by immersion of the microstructure for 2 hours in a 1:3 mixture of 30% (v/v) H$_2$O$_2$ and H$_2$SO$_4$. After thoroughly rinsing with

Table 5.4: Anisotropic reactive ion etching parameters. The following etching parameter were repeated 100 times at a temperature of -10°C in order to obtain etching depths of 10 μm.

	t (sec)	p (mTorr)	CHF$_3$ (sccm)	SF$_6$ (sccm)	P$_{RF}$ (W)	P$_{ICP}$ (W)
etching	8 sec	50	0	20	30	90
passivation	5 sec	18	15	40	25	300

deionized water, the structures were blown dry with nitrogen and ready for the following nano-patterning step.

5.3.2 Nano-patterning of silicon microstructures

Silicon microstructures were decorated with patterns of nano-sized gold particles by using BCMN [53]. The diblock copolymer solution was prepared by dissolving 102 mg of the poly(styrene)-block-poly(2-vinylpiridine) (PS-b-PVP) (P4988-S2VP; Polymer Source Inc., Canada) with a molecular weight of PS(110000)-b-PVP(52000) in 20.4 mL ortho-Xylol (Merck, Germany) and stirred over night at r.t.. By adding 30.6 mg tetrachloroauric acid (HAuCl$_4$) (Sigma, Germany) to the micellar polymer solution, the micelles were loaded with the auric acid at a molar loading ratio of 0.25. The solution was stirred over night at r.t. in order to obtain a high monodispersity of the micells' size. Then, 100 μL of the micellar solution were dropped on the microstructured silicon wafers while excessive solution was removed with a tissue at the edge of the wafer and the solution was let to completely evaporate. In order to reduce the auric acid to elemental gold and remove the diblock-copolymer, the substrates were treated for 45 min with a hydrogen/argon (1:9) plasma with a pressure of 0.4 mbar and a power of 150 W (100-E; PVA-TePla, Germany).

5.3.3 Replica molding of PEG-hydrogel

PEG diacrylate (Mw = 700, 6 mL) was mixed with 390 μL of a freshly prepared saturated aqueous solution (7.6 mg/mL) of the photo-initiator 2-hydroxy-4'-(2-hydroxyethoxy)-2-methylpropiophenone (#410896; Aldrich, Germany), degassed and stirred under nitrogen atmosphere. A casting chamber with the structured silicon wafer on the bottom and a quarz glass on top, separated by a 4 mm thick spacer was filled with the PEG solution and irradiated under UV-light for 300 sec with an intensity of 4.5 W/cm^2 (Lightning cure LC8; Hamamatsu, Japan) in order to initiate the radical polymerization. Swelling of the hydrogels in deionized water resulted in detachment from the casting chamber.

5.4 Material characterization techniques

5.4.1 White-light interferometry

Height profiles of the developed photoresist patterns on the wafers were measured with a white-light interferometer (NewView 5000; ZygoLOT, Germany) and the raw data was analyzed with the MetroPro software (V 7.10.0; ZygoLOT, Germany).

5.4.2 Scanning electron microscopy

For scanning electron microscopy imaging, PDMS replicas were coated with a 50 nm gold layer in a sputter coater (0.5 bar, 60 mA, 60 sec; BalTec MCS 010; Leica, Germany) to reduce charging effects of the polymer during imaging. To image the microstructure of the PDMS replica, a field emission scanning electron microscope was used with a Schottky cathode (Zeiss Ultra 55, Carl Zeiss, Germany) with an acceleration voltage of 3–10 kV, at a pressure of $< 5 \cdot 10^{-6}$ mbar, a beam diameter < 5 nm and the InLens detector.

5.5 Cell experiments

5.5.1 Cell culture

Human pancreatic epithelial cancer cells (Panc-1) (European collection of cell cultures, UK), and Panc-1 stably transfected with keratin K8-eCFP/K18-eYFP (P1) [95], kindly provided by Professor Seufferlein (Martin-Luther-University Halle-Wittenberg, Halle), were cultured in Dulbecco's modified Eagle's medium (DMEM) (#10938; Life Technologies, Germany) and supplemented with 10% FBS (#A11-151; PAA Laboratories, Germany), 1% L-glutamine (#25030; Life Technologies, Germany) and 1% penicillin-streptomycin (#15140148; Life Technologies, Germany). The medium for the P1 cells contained in addition 0.03% hygromycin B (#ant-hg-1; Invivogen, France) and 1% gentamycin (#47991; Serva electroporesis, Germany) as selection agents for the transfected cells. For subculturing and prior to experiments, cells were detached with 0.25% trypsin-ethylenediaminetetraacetic acid (EDTA) (#T4049; Sigma). The keratin morphology of adherent cells was changed by adding sphingosylphosphorylcholine (SPC) (Merck, Germany) from a 10 mM aqueous stock solution to a final concentration of 10 μM. Diameters of suspended cells were automatically measured with a Z2 Coulter Counter (Beckmann Coulter, Germany) via an electrical sensing zone method.

5.5.2 Cell migration and live-cell imaging

Prior to the experiments, cells were resuspended in medium containing 5% FBS with a final concentration of $1-2*10^6$ cells/mL. Approximately 100 μL of this cell suspension was gently introduced into the chip to seed the cells in close proximity to the channels. Image capturing for time-lapse videos was started 30 min after cell seeding for non-SPC treated cells or otherwise 60 min after addition of SPC. Phase-contrast live-cell imaging was performed in a heated and air-humidified chamber built around an automated inverted microscope (Axio Observer.Z1, EC PlanNeofluar 10x/0.3 Ph1; Carl Zeiss, Germany) and controlled with the Axio-

Vision software (AxioVision V.4.6; Carl Zeiss, Germany). The motorized stage enabled the observation of up to 320 channels in one experiment. For time-lapse videos, images were taken every four minutes at each position over a period of 16 h. Confocal live-cell images of P1 cells were taken with a spinning disk setup UltraVIEW ERS LiveCell Imaging scanner (PerkinElmer, USA) connected to an inverted microscope (Axio Observer.Z1, LD C-Apochromat 40x/1.1W Korr UV-VIS-IR, Carl Zeiss, Germany), an argon/krypton laser (488/568/647 nm; Melles Griot, USA) and controlled with the Volocity 3.6 software (PerkinElmer, USA).

5.5.3 Immunocytochemistry

Filamentous actin, keratin and nuclei of fixed Panc-1 cells were fluorescently labeled and imaged under the microscope; for the detailed steps of the staining protocol see Table 5.5. Cells were fixed with 4% para-fomaldehyde (PFA) (#P6148; Sigma, Germany) in PBS and permeabilized with 0.5% Triton-X100 (#T8787; Sigma, Germany) in PBS and blocked with fish skin gelatine (#G7765; Sigma, Germany). Keratin localization was detected with the monoclonal pan-cytokeratin (KL1) antibody (#IMG-80127; Imgenex, San Diego, CA) in combination with Alexa488 chicken anti mouse (#A-21200; Life Technologies, Germany) antibody; filamentous actin was visualized with AlexaFluor568 phalloidin (#A12380; Life Technologies, Germany) and nuclei were stained with Hoechst 33342 (#H3570; Life Technologies, Germany). For fluorescent imaging, coverslips with fixed cells were mounted on coverslides using Aqua-Poly/Mount (#18606; Polysciences, Germany). Images were acquired with an AxioCam MRm3 charged coupled device (CCD)-camera on an Axio Observer.Z1 inverted microscope equipped with a Plan-Apochromat 63x/1.4Oil objective and a Colibri light emitting diode (LED) illumination system (all Carl Zeiss, Germany).

5.6 Image analysis and data processing

Images were processed and analyzed using the ImageJ software [96]. The manual tracking plug-in was used to follow migration of the cells inside the channels on flat

Table 5.5: Staining protocol for Panc-1 cells. Labeling of keratin, filamentous actin and nuclei.

wash	1x		37°C	PBS
fix	1x	15 min	4°C	4% PFA in PBS
wash	2x	2 min	r.t.	PBS
permeabilize	1x	10 min	r.t.	0.5% Triton X-100 in PBS
wash	3x	2 min	r.t.	PBS
block	1x	10 min	r.t.	0.2% gelatine in PBS
first antibody	1x	over night	4°C	mouse α-cytokeratin 1:200 in 0.5% Triton X-100/0.2% gelatine in PBS
wash	6x	20 min	r.t.	PBS
second antibody	1x	2 h	r.t.	AlexaFluor 488 chicken α-mouse IgG 1:1000; AlexaFluor 568 phallodin 1:50; Hoechst 33342 0.2 µg/mL in 0.5% Triton X-100/0.2% gelatine in PBS
wash	6x	20 min	r.t.	PBS

surfaces or on adhesion-mediating lines. The data obtained from the cell tracking were processed using routines written with MATLAB (Version 7.5; The MathWorks, USA). To determine the nuclei diameters, cells were stained with 0.2 µg/mL Hoechst 33342 and average diameters were calculated by taking the projected fluorescent areas (A) of the cells and simply calculating the corresponding diameters (d) of a perfectly spherical shape with the geometric equation $d = \sqrt{\frac{A}{\pi}} * 2$.

In order to quantify the different behaviors of the cells that contacted the channels, each instance was assigned to one of the following three different categories. (i) Cells that penetrated the channels with their cytoplasm to a depth of at least 20 µm were classified as *penetrating* cells. This minimum limit of penetration depth was chosen in order to exclude cells that were moving perpendicular to the channel direction without a change in their directionality (8% of the total number of interacting cells for the -SPC and 12% for the +SPC condition). (ii) All the cells that completely entered the channel structure and then stopped migrating or turned around were called *invasive* cells. (iii) Cells that migrated completely to the other side of the channel were termed *permeative* cells. Finally, cells that were still migrating through the channels when the image acquisition stopped after 16 hours (13% for -SPC and 27% of the cells for +SPC) were excluded from classification, as it was not possible to determine their assignment between categories (ii) and (iii).

The cells' mean migration speeds were measured by tracking the displacement of the moving cell from frame to frame in a time-lapse experiment. The cell's mean speed was then calculated from the obtained Euclidean distances over a time period of at least 3 h. Analysis of the dynamical change in cell length of cells migrating inside the channels and on adhesive lines was carried out by plotting the cell length over time and calculating its mean value within comparable time frames. The mean cell length was set to one and the respective normalized standard deviations (s.d.s) were plotted over the time. Low values of the normalized s.d.s refer to low fluctuations in the cell length (smooth *sliding*) while higher values represent higher fluctuations (*push-and-pull*-like pattern).

Statistical analysis was carried out using MATLAB and Kaleidagraph (Version 4.0, Synergy Software, USA). Errors are given as standard error of means (s.e.m.) if not differently indicated.

Chapter

6

Results

6.1 Micro-fabricated migration chip

A micro-fabricated device was developed with channel structures mimicking a confined 3D environment. The key feature are micro-sized channels of different lengths (50-300 μm) and widths (3-20 μm) that are connecting two reservoirs (Figure 6.1 A). Micro-sized structures were fabricated using a two-step photolithography pro-

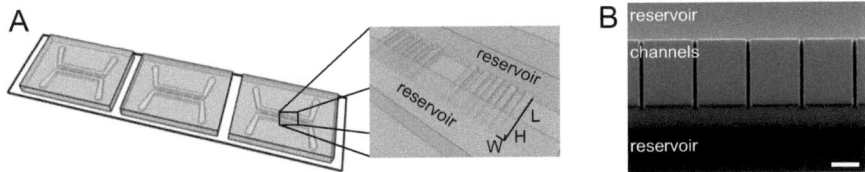

Figure 6.1: Micro-fabricated channel structures. (A) Schematic of three migration chips assembled on a glass slide with a close-up of two reservoirs connected by micro-sized channels. (B) Scanning electron microscope image of the cast PDMS-based replica showing 7x11x150 μm micro-grooves on a 150 μm high plateau that separates the two reservoirs in the assembled migration chip. Scale bar, 50 μm.

cess with the mechanically stable SU-8 photoresist that can be used to pattern high aspect-ratio structures. With the two-step lithography process microstructures of two different heights are built up. The first layer of photoresist determines the channel height and width, while the second layer determines the channel length and the height of the reservoirs. A rotationally symmetric layout of the photomasks was chosen and by rotation of the first and second photomask by 90° with respect to each other during the exposure, the number of possible combinations of channel widths and lengths was doubled. Figure 6.2 shows examples of two fabricated wafers that serve as casting molds for the following replica-molding process. Cross-

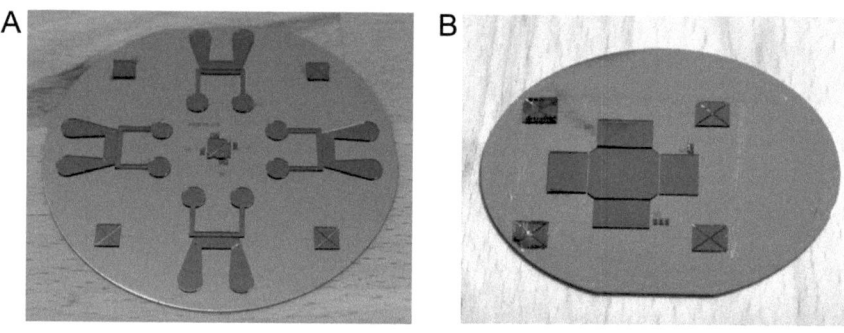

Figure 6.2: Two different designs of master substrates for migration chips. (A) Silicon wafer patterned with two layers of SU-8 resist ready for replica molding. Four migration chip chambers are arranged in a rotationally symmetric way on the wafer. The four squares are alignment markers that are needed for the correct positioning of the second photomask. The wafer diameter is thee inches. (B) Silicon wafer patterned with two layers of SU-8 resist used for replica molding. The mold is for one square-shaped migration chip where 80 channels are located along each of the four sides. The four small squares are alignment markers that are needed for the correct positioning of the photomask in the second exposure step. The wafer diameter is two inches.

linking the silicon based PDMS prepolymer on these master structures leads to a transparent elastic polymer mold that was gently removed and covalently bound to a glass slide as shown in Figure 6.3 B. Two different designs of these migration chips were developed where the microchannels were either integrated into a closed microfluidic setup where the two reservoirs could be separately filled (Figure 6.3 A and B) or an easy-to-use open setup with more channels available in each experiment (Figure 6.3 C) where the cell suspension is directly injected in close proximity to the channel entrances with a normal pipette. Cell invasion and

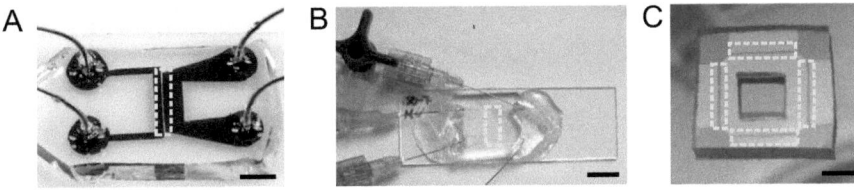

Figure 6.3: Assembled migration chips. (A) Migration chip with connected tubes for liquid handling and filled with blue ink for better visualization of the reservoirs. Scale bar, 5 mm. (B) Assembled migration chip with connected needles for liquid handling, ready for cell experiments. Scale bar, 10 mm. (C) Example of assembled migration chip that can be easily used for cell experiments by directly injecting the cell suspension into the middle of the chip. Regions where the microchannels are located are outlined with yellow dashed lines. Scale bar, 5 mm.

migration through these precisely defined channel structures was monitored under standard cell culture conditions in real-time using an automatically controlled microscope with transmitted light or fluorescence live-cell imaging. Examples of such time-lapse videos are shown in Videos A.1.1-A.1.3.

6.2 Migration studies of human pancreatic epithelial cancer cells

6.2.1 General cell behavior inside the migration chips

Human pancreatic epithelial cancer cells (Panc-1) seeded in the migration chips showed their typical spreading morphology and migration behavior on the flat surface in front of the channels where they migrated in a non-directed way. However, as soon as the cells reached the walls between the single channels, they started to preferentially move along the walls, a phenomenon known as "contact guidance" [55,69]. Once the cells approached the entrance of a microchannel, they were able to migrate through it as long as the channel cross-sections were relatively large (WxHxL: 15x11x50 μm) as it is shown in Video A.1.1. This permeative migration behavior through the channels did not have a major effect on the migration speed of the cells nor were the cells forced to deform themselves in a substantial way. However, by reducing the channel width from 15 to 7 μm, only 7% of the Panc-1 cells that initiated contact with the channels were able to permeate them. Moreover, they had to deform themselves in a dramatic way to enter the channels (Figure 6.4 A, B). At channel widths of 3 μm cells were not able to invade the channels although parts of their cytoplasm extended into it as shown in Figure 6.4 C.

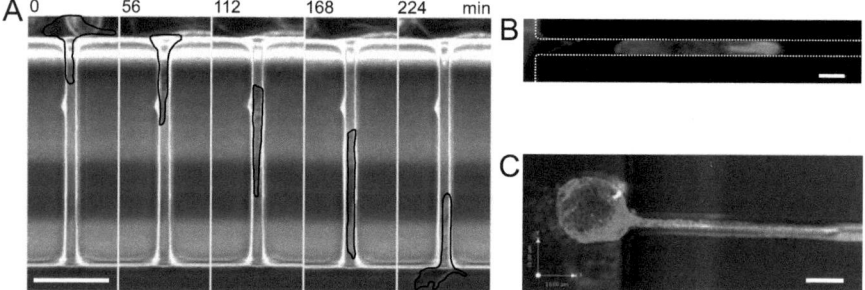

Figure 6.4: Interaction of cells with channel structures. (A) Images of a time-lapse video with a SPC-treated Panc-1 cell permeating a channel (7x11x150 μm). For better visualization, the cell was encircled manually. Timesteps are indicated in each image; scale bar, 50 μm. For the whole sequence see Video A.1.2. (B) Panc-1 cell inside a channel (7x11x150 μm) migrating from left to right with labeled cytokeratin (green), actin (magenta) and nuclei (blue), non-SPC-treated. Scale bar, 10 μm. (C) Scanning disk live-cell image of a P1 cell with fluorescently labeled keratin 8 and 18 (green) penetrating a channel (3x11x150 μm). The cell membrane inside the narrow channel is highly deformed while the spherical nucleus is located outside at the channel's entrance. Scale bar, 10 μm.

In order to estimate the volume that a cell occupies inside the channels, the corresponding cell volume was calculated from the diameter of cells in suspension that was measured to be 19±2 μm. Thus, a cell migrating inside a 7 μm wide, 11 μm high and 150 μm long channel will occupy one third of the total channel volume. Assuming that the cell volume remains constant for the cells in suspension and inside the channels, the change in geometry from a sphere in solution to a rectangle, caused by the confined geometry of the channel, leads to a 1.6 fold increase in total surface area of the cell.

6.2.2 SPC alters spatial keratin organization, 2D cell migration, and invasive cell behavior

Treatment of Panc-1 cells with SPC leads to a change in the organization and distribution of the keratin filaments inside the cell. Under normal cell culture conditions, keratin filaments are distributed all over the cytoplasm with a slightly higher concentration around the nucleus as it can be seen in the fluorescently labeled cells in Figure 6.5. After incubation of the cells with a 10 μM SPC solution

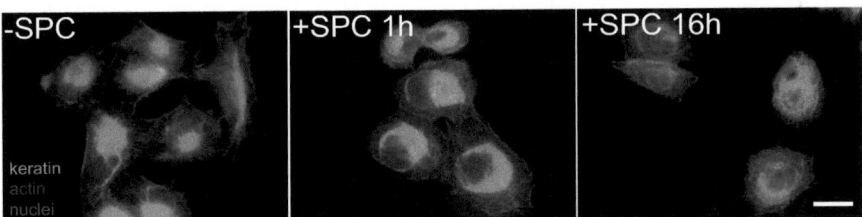

Figure 6.5: SPC-effect on cytoskeleton structure. Panc-1 cells without SPC treatment (-SPC) and after 1 hour and 16 hours of incubation with 10 μM SPC (+SPC). Fluorescently labeled cytokeratin (green), filamentous actin (magenta) and the nuclei (blue). Scale bar, 50 μm.

for 1 hour, the keratin filaments get highly compacted and concentrated around the nucleus, a reorganization that can be still observed, although less drastic, after an incubation time with SPC of 16 hours (Figure 6.5). At the same time, the SPC treatment also increases the motility of Panc-1 cells on a flat surface. However, not all of the cells respond tot he SPC treatment as quantified by optically examining the keratin organization of immunostained cells. In the absence of SPC, about 33% of the Panc-1 cells exhibited a predominantly perinuclear organization of keratin 8/18 filaments. This number increased to about 66% in the presence of the bioactive lipid which is in accordance with previous data [91].

Migration speeds of cells on flat surfaces were measured with cells that migrated on the flat areas inside the migration chip. The non-treated cells migrated at speeds of 0.32±0.04 μm/min (N = 195), while upon SPC treatment this value was significantly increased by a factor of 1.5, leading to migration speeds of 0.47±0.1 μm/min (N = 252). More details of the migration dynamics will be discussed in section 6.2.3.

Cells inside the channels are highly elongated as shown in Figure 6.4. The organization of the cell's keratin and actin network as well as the elongated nucleus inside the channels can be seen for a fixed and fluorescently labeled cell in Figure 6.4 B and the keratin dynamics of a P1 cell inside a channels is shown in Video A.1.3.

In order to describe the behavior of the cells that initiated contact with the channels, they were assigned to one of the following three categories: (i) *penetrating* cells, (ii) *invasive* cells and (iii) *permeative* cells as described in section 5.6. An example of a cell permeating a microchannel is shown in Figure 6.4 A and examples of all three different behaviors can be seen in Video A.1.2. The categorized behavior of cells that interacted with the microchannels is summarized in Figure 6.6.

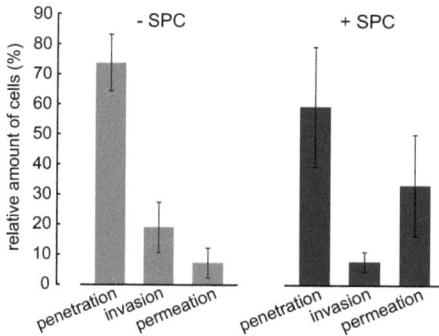

Figure 6.6: Summary of Panc-1 cell interactions with the channels. Non-SPC-treated cells (-SPC, 4 experiments, N = 100) and cells treated with 10 μM SPC (+SPC, 4 experiments, N = 115).

In the absence of SPC, the amount of cells that *invaded* or *permeated* the channels was relatively low, 19% and 7%, respectively, as compared to the 73% of cells that only *penetrated* the channels while their nuclear region stayed outside the channels. This ratio changed significantly once the cells were treated with SPC. The addition of SPC resulted in a highly increased *permeative* behavior of the cells. In response to SPC treatment 33% of the cells *permeated* the channels, which is a 4.7-fold increase over the *permeative* behavior of non-SPC-treated cells. This difference in cell behavior is remarkable, given that not 100% of the cells respond to SPC treatment as described above.

To elucidate whether the nucleus size played a role in determining the cell's *invasive* behavior, the nuclei diameters of cells growing on a flat surface were measured, with respect to SPC treatment. The non SPC-treated cells had nuclei diameters of 17.0±0.4 μm (N = 77), and cells exposed to SPC for 1 hour and 4 hours had nuclei diameters of 14.9±0.17 μm (N = 196) and 15.9±0.27 μm (N = 75), respectively. Thus, the nuclei diameters were reduced by 12% after 1 hour of SPC treatment. Nevertheless, the compacted nuclei with a diameters of 14.9±0.17 μm were still too big to fit into channels with a cross-section of 7x11 μm without being deformed.

6.2.3 3D channel structures have an impact on migration dynamics

Cell migration of Panc-1 cells was analyzed on flat 2D surfaces (Figure 6.7 A) and compared with the one inside 3D channels (Figure 6.7 B). In order to elucidate whether the 3D architecture of the channel walls or only the restriction in directionality by the channels changed the migration phenotype of the cells, migration behavior of Panc-1 was additionally monitored along 1D adhesive lines (Figure 6.7 C and Video A.1.4).

Comparing the 2D migration speed of non-SPC-treated cells (0.32±0.04 μm/min) with the migration speed inside the channels, an approximately three-fold increase was observed for both, non-SPC-treated (1.06±0.1 μm/min; N = 13) and SPC-treated cells (1.19±0.1 μm/min; N = 26). However, migration speed inside the channels was not significantly increased upon SPC treatment with speeds of 1.06±0.1 μm/min before and 1.19±0.1 μm/min after SPC treatment.

Cell migration speed along the lines was slightly higher than inside the channels (1.2±0.19 μm/min; N = 12) and also not significantly influenced by SPC treatment (1.15±0.01 μm/min; N = 13). The different migration speeds are summarized in Figure 6.7 D.

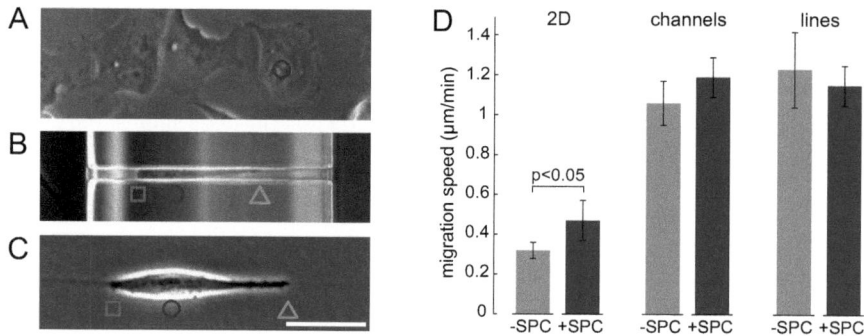

Figure 6.7: Migration speeds in different environments. (A–C) Examples of cells migrating on a flat surface (A), inside a channel (B), and along an adhesion mediating line (C). All cells are migrating from left to right and the position of the front (green triangle), center (blue circle) and rear (red square) are indicated; scale bar, 50 μm. (D) Comparison of the migration speed of cells migrating either on a flat 2D glass surface, through microchannels or along adhesion mediating lines with respect to the presence of SPC. The average speed of Panc-1 cells on a flat glass surface 2D is increased upon treatment with SPC. Migration speed in microchannels is generally higher but not affected by SPC treatment; that is also the case for migration on lines. Statistics: student's t-test, p-value as indicated; error bars show s.e.m..

To investigate whether the tumor cells show a coordinated and characteristic movement, leading edge, nucleus, and rear edge of *permeating* cells and cells on adhesive lines were separately tracked, as indicated in Figure 6.7 B and C. Plotting the absolute position inside the channel versus time, two characteristic patterns of motion can be distinguished: a smooth *sliding* motion and a stepwise *push-and-pull* behavior (Figure 6.8 A and B, respectively).

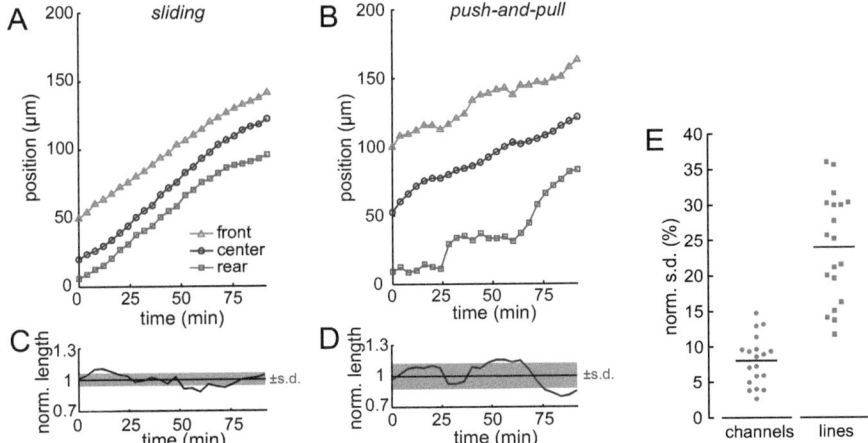

Figure 6.8: Cell migration dynamics inside channels and on lines. (A) Typical profile of a *sliding*-like migration pattern inside a channel with a width of 7 μm where the cell's front, center, and rear are moving in a synchronous, almost equidistant manner. (B) Typical profile of a *push-and-pull*-like migration behavior of a cell on an adhesion mediating line with a width of 7 μm. The distance between between the cell's front rear changes in an oscillating manner independent of the nucleus displacement. (C, D) Length variation of individual cells during migration, normalized to the mean cell length (black line). Standard deviations (s.d.) of the mean length are indicated by the shaded rectangle. *Sliding*-like migration is characterized by smaller s.d. (C) than the ones obtained for *push-and-pull*-like patterns (D). (E) Distribution of normalized s.d.s of single cells performing a directed migration inside channels or along lines. The lines indicate mean values.

The *sliding* pattern is characterized by a quasi-equidistant movement of the front and rear of the cell. To quantify the quasi-equidistant movement, the s.d. of the normalized mean cell length was calculated; an example of variation in cell length normalized to its mean length and the corresponding s.d. is given in Figure 6.8 C. For the cells inside channels this normalized s.d. was found to be 8% ($N = 19$); see Figure 6.8 E.

In contrast, the *push-and-pull* migration pattern is characterized by a variation of the cell length in an oscillating manner (Figure 6.8 D), similar to the classical stepwise migration pattern of fibroblasts on a flat surface as described in section 2.1 and [70]. This leads to an increased normalized s.d. of the mean cell length with a value of 24% ($N = 19$) for the cells migrating on adhesive lines see Figure 6.8 E.

6.3 Invasiveness of medulloblastoma cells

In collaboration with Stefan Pfister and Alfa Bai at the German Cancer Research Center (DKFZ) in Heidelberg, Germany, the migration chips were used to test the migratory and invasive potential of cells that originate from medulloblastoma (MB), a highly malignant primary brain tumor that originates in the cerebellum or posterior fossa [97].

In molecular biology and medical research, the identification of micro ribonucleic acids (miRNAs) as mediators of various centrally important cellular processes has been an important improvement in understanding molecular cell functions within the last years. These short, non-coding ribonucleic acid (RNA) molecules (21-25 nucleotides) are known to play important roles in tumorigenesis and to display distinct expression signatures within different cancer types [98, 99]. However, the specific roles of miRNAs in the biology of MB, the most frequent malignant brain tumor in childhood remain poorly understood.

The MB cell lines DAOY (#HTB-186, American Type Culture Collection, USA) and Med8A were selected for functional validation according to their different expression levels of candidate miRNAs. The miRNA miR-182 is a potential cancer associated miRNA, a so called oncomir, of MB subtype which is triggered by the non-sonic hedgehog pathway [100]. The DAOY cell line is characterized by a high native expression level of miR-182, compared to the rather low levels of miR-182 in the Med8A cell line [101].

Stable overexpression or transient knockdown of miR-182 and miR-183 in the MB cell lines DAOY or Med8A did not consistently affect proliferation, but significantly changed their migratory and invasive behavior as tested with conventional scratch assays [101]. However, it was not known in how far different expression levels of miR-182 change migration dynamics and invasive behavior on a single-cell level.

In order to further examine the migration dynamics, the developed migration chip was used to monitor and quantify changes in migratory and invasive behavior of DAOY and Med8A cells in more detail [101]. For the migration studies of MB cells, migration chips were employed with channel dimensions of 5x11x300 μm as this was the channel cross-section where the MB cells showed the highest variations in invasive behavior depending on the varied miRNA expression levels.

The image sequence in Figure 6.9 A shows an example of a DAOY cell permeating a microchannel with the dimensions of 5x11x300 µm. Two clones of sta-

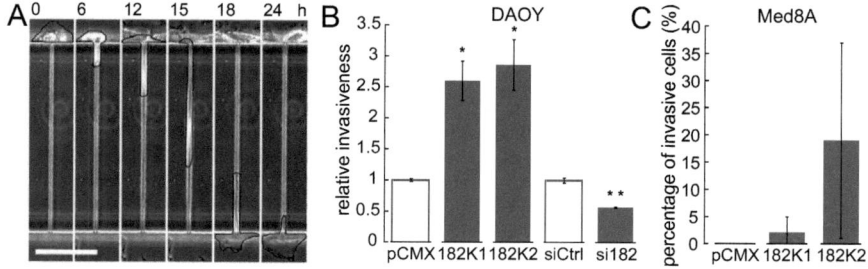

Figure 6.9: Invasive behaviors of MB cells. (A) Image sequence from a time-lapse video showing a DAOY cell permeating a 5x11x300 µm channel. The outline of the DAOY cell was manually marked with black lines for better visualization; scale bar, 100 µm. (B) Relative invasiveness of DAOY cells when either stably overexpressing miR-182 (182K1 and 182K2) or after transient knockdown with siRNA treatment (si182) compared to empty vector transfected control (pCMX) and scramble negative siRNA treatment control (siCtrl). Statistics: Dunnett contrasts with *: $p < 0.001$ and **: $p = 0.03$. (C) Percentage of invasive Med8A cells, stably transfected with empty vector (pCMX), or overexpressing miR-182 (182K1, 182K2). All bars indicate the mean values ±s.d. of three independent experiments. Figure adapted from [101].

bly transfected DAOYcells, overexpressing miR-182 (182K1 and 182K2) as well as DAOY cells, with transientelly knock-down of mi-R182 with siRNA, were tested regarding their invasive behavior with the results shown in Figure 6.9 B: Stable overexpression of miR-182 increased the average invasiveness 2.6-fold (DAOY-182K1) and 2.9-fold (DAOY-182K2) compared to DAOY cells with the empty pCMX vector (DAOY-pCMX), respectively. Inversely, DAOY with transient knockdown of miR-182 by small interfering ribonucleic acid (siRNA) treatment showed a significantly reduced invasiveness of 0.6 fold as compared to negative controls with empty vector (pCMX) and scramble negative siRNA (siCtrl). Examples of the differently treated DAOY cells are shown in Videos A.2.1-A.2.4

Two stably transfected clones of Med8A cells, overexpresing miR-182 (182K1 and 182K2) were also tested regarding their invase behavior into microchannels of 5x11x300 μm. In contrast to the DAOY cells, Med8A cells transfected with an empty vector (pCMX) showed no migratory behavior and were not capable of invading the 5 μm wide channels. However, after miR-182 overexpression, Med8A cells became slightly migratory and a considerable proportion of invasive cells were observed, (Figure 6.9 and Video A.2.5 and A.2.6), consistent with the findings in DAOY cells.

Supported by these findings it could be further demonstrated in additional *in vitro* and *in vivo* experiments that overexpression of miR-182 drives metastatic dissemination and that inhibition of miR-182 is a molecular target for the treatment of metastatic MB [101].

6.4 Micro-sized hydrogel channels with nano-patterned walls

Structuring materials at different length scales is an important and demanding challenge, in particular for studying biological systems, but also in the field of materials science in general. Especially, at the interphase where synthetic materials get in contact with living cells, it is of high importance to structure and biofunctionalize materials. Structuring and functionalization methods, typically aim for specifically mimicking characteristics of the ECM that control and influence the cell behavior.

In this section an approach is introduced which allows for the fabrication of a PEG-based hydrogel with a microstructured topography that is decorated on its surface with gold nanoparticles. Such a complex composite material can be tuned in the following ways: First, the stiffness of the hydrogel can be adjusted via different monomer lengths and cross-linking densities, to adjust the gel to different stiffnesses that correspond to different stiffnesses of the ECM. Second, as the gel is polymerized in a microstructured casting mold, a variety of micro-sized gel topography can be obtained, that may mimic scaffolds of the ECM in the micrometer range. Finally, the gold nanoparticles provide a chemical contrast to the remaining surface of the hydrogel, which allows for selectively functionalizing the gold particles, for example with specific ligands of the ECM while the hydrogel remains protein repellant. Additionally, the density and size of the gold particles can be adjusted, which allows for the control of concentration and distance between single ligands presented on the surface.

The results of the fabrication steps are presented and discussed in the the following. The photo-reversal photoresist AZ 5214E, used as a negative tone photo resist, was found to be well suited to passivate the surface of a silicon wafer that were anisotropically etched using a RIE process with gas mixtures of SF_6 and CHF_3. The AZ 5214E processing was adjusted to form a 1.5-1.8 μm thin layer at the irradiated regions with a slightly undercut profile as shown in the scanning electron microscopy (SEM) image in Figure 6.10 A.

Figure 6.10 B and C show etched microstructures with the passivating photoresist still covering the silicon. The microstructures were etched between 7 and 10 μm deep, and as seen in Figure 6.10 C ablation of the resist during the etching resulted in a reduced layer thickness of 1.1 μm.

Figure 6.10: Photoresist on silicon wafer before and after RIE. (A) Profile view of developed AZ 5214E photoresist with a thickness of 1.7 μm and an undercut profile. Scale bar, 1 μm. (B) Profile view of an anisotropically etched silicon wafer with a protective photoresist layer above the non-etched area. Scale bar, 5 μm. (C) Tilted view on 3 μm thick line, covered with photoresist. Scale bar, 5 μm.

Once the remaining photoresist was dissolved, the final microstructures, comprising linear features with a width of 3-10 μm and a height of 10 μm, were obtained on the etched silicon (Figure 6.11 A and B), while only sparsely distributed nano-sized pillars were found as imperfections on the etched surface.

In the following step, the obtained microstructured silicon mold was patterned with gold nanoparticles by means of BCMN. Typically, BCMN is used to pattern large areas (hundreds of square micrometers) of flat silicon or glass surfaces with hexagonally ordered nanoparticles (e.g. gold-spheres with a diameter of 5-10 nm) with an inter-particle distance adjustable between roughly 20 to 200 nm. The high order and the variation in inter-particle distance is obtained through a controlled dewetting of a micellar solution in an organic solvent (dip-coating).

As for flat surfaces the dewetting of the micellar solution can take place in a very homogeneous way, this was not the case for microstructured silicon substrates where capillary forces at the corners of the walls lead to an accumulation of the micellar solution while the top of thin microstructures dewetted even faster

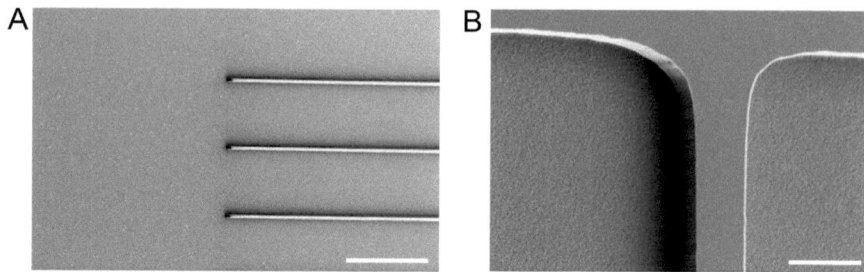

Figure 6.11: Microstructured silicon wafers. (A) Tilted view on 7 μm thick and 10 μm high lines on etched silicon. Scale bar, 100 μm. (B) Tilted view on 3 μm thick and 10 μm high lines on etched silicon. Scale bar, 100 μm.

than plain surfaces. Therefore, the dewetting technique was modified and the best results were obtained by directly spotting the micellar solution on the structured wafer with a glass pipette while ensuring a fast dewetting by soaking up the excessive solution with a clean tissue.

Examples of the BCMN process applied on microstructured silicon wafers are shown in Figure 6.12. Typically, the nano-patterning with the gold particles was incomplete on the top of the micro-sized silicon walls, while on the bottom corners high gold particle concentrations, so called "multi-dots" were formed.

The last step in the fabrication of the nano-patterned microstructures was the casting of the PEG hydrogel on the microstructured silicon mold while simultaneously transferring the gold particles onto the hydrogel via a UV light initiated radical poymerization reaction as reported elsewhere [53]. Figure 6.13 shows photographs of the structured silicon mold, the mold with a PDMS cage that can be filled with the PEG monomer solution, and the final polymerized PEG hydrogel with the microstructures.

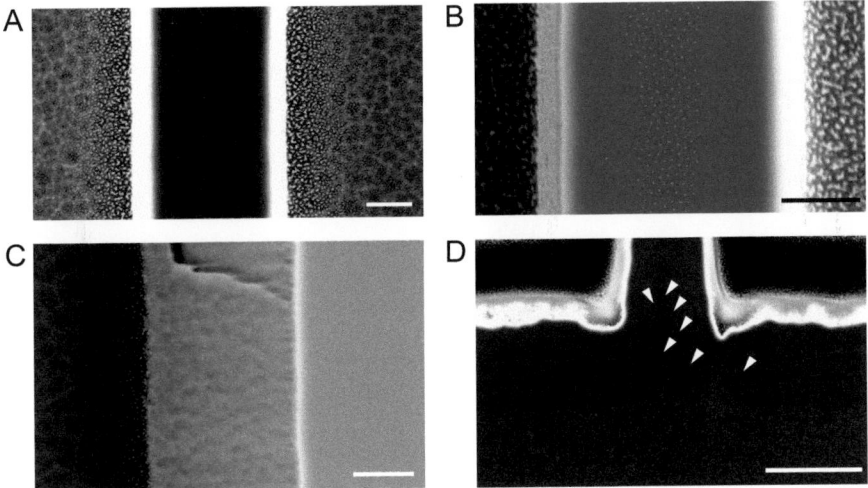

Figure 6.12: Silicon microstructures patterned with gold nanoparticles. (A) Multi-dots found at the left and right bottom of a 3 μm wide wall. Scale bar, 1 μm. (B) Sparsely distributed gold nano particles on top of a 3 μm thick line. Scale bar, 1 μm. (C) Tilted view on a 10 μm high wall with gold nanoparticles homogeneously distributed on the side-wall. Scale bar, 1 μm. (D) Gold nanoparticles on the top plane of a microstructured silicon wafer. Sparsely distributed gold particles at the beginning of a 3 μm thick wall are indicated with white arrows. Scale bar, 5 μm.

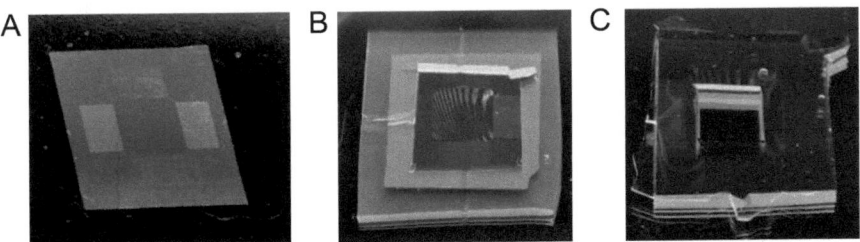

Figure 6.13: Microstructured silicon wafer and hydrogel casting. (A) Reactive ion etched silicon wafer with microstructures. (B) PDMS barrier for the PEG casting step on a structured silicon wafer. (C) Polymerized microstructured PEG hydrogel.

Chapter 7

Discussion

7.1 Single cell migration through micro-sized channels

7.1.1 Invasiveness and migration dynamics of Panc-1 cells

With the developed migration chip, channel structures of sub-cellular dimensions have been applied to study cell deformation and invasiveness as well as migration dynamics and coordination of cellular movement. The setup tailored for transmitted light or fluorescence live-cell imaging applications. Due to its precise channel architecture, it allows for a quantitative analysis of mechanical deformation for cells migrating through confined 3D environments. In the last few years, similar channel architectures for studying leukocyte migration inside a confined 3D environment have been used [102, 103]. Recently, Irimia *et al.* [104] examined the migrational persistence of cancer cells inside micro-sized channels, demonstrating the usefulness of such single-cell based approaches.

Using the functionality of the migration chip, the question was addressed, whether the known effect of SPC on Panc-1 cells (decreasing the cells' stiffness) also enables the cell-driven invasion into confined spaces. Beil *et al.* measured a drop in the Panc-1 cells' elastic modulus from about 28 to 16 mN/m upon SPC treatment using a microplate-based single-cell stretcher [91]. Using a Boyden chamber assay they correlated this decrease in stiffness to an increased ability of the cells to squeeze through the membranous pores. In the present study, the behavior of cells initiating contact with channels of a cross-section of 7x11 μm, a size at which the Panc-1 cells were hardly able to squeeze through, was characterized and quantified. Employing the precisely confined channel architectures, on the one side the results of Beil *et al.* were confirmed, as a five-fold increase in the number of cells permeating the channels upon SPC treatment was observed and on the other side the migration dynamics inside the channels could be characterized and quantified.

The enhanced invasive behavior upon treatment with SPC may be explained by using a simple two-component model. First, there is the motor unit with the driving actin-polymerization (lamellipod) in the front of the cell and the contractive acto-myosin assembly at the rear. Secondly, there is the rather passive and voluminous cell body being pulled upon migration (Figure 7.1). For an *invasion*

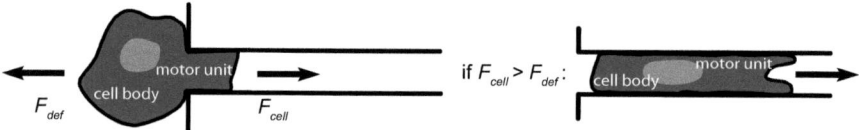

Figure 7.1: Schematic of a simple two-component model describing cell invasion.

into the channel, the cell needs to deform the cell body requiring compression of the keratin network and the nuclear region. Inside the channels, nucleus shape and keratin network structure deviate from the normal spatial distribution in the cell as shown in Figure 6.4 B. Assuming that Panc-1 cells are able to generate only a finite force, it can be speculated that the motor unit of a non-SPC-treated Panc-1 cell is not able to pull the cell body into a channel with a width of 7 μm. This speculation is supported by observations on neutrophil leukocytes migrating inside small capillaries. These cells are able to counterbalance a maximal hydrostatic counter-pressure of 1.5 kPa by generating a traction force of 38 nN inside the capillaries [105]. Additionally, it was shown for leukocytes that the spatial organization of intermediate filaments plays a major role for the cells' migratory behavior [106].

Beyond elucidating the effects of size-exclusion on the invasive behavior of cells, the migration chip permitted the quantification of drastic differences in 2D and 3D migration speeds. This property is of particular importance since recent studies attributed increases in migration speed to the dimensionality of the cell environment [32, 107]. The observed increase in 2D speed after SPC treatment might be attributed either to the keratin reorganization and softening of the cell or to the known enhancement of filamentous actin formation by SPC [90]. As the speed inside the channels is not affected by SPC treatment it seems reasonable to speculate that the SPC-effect on actin may not be the major factor, but that the different keratin morphology might cause the increased migration speed on flat surfaces. Assuming that the nuclear region and keratin envelope of a cell inside a channel is in a compacted state, its steric hindrance should have only a minor influence on the migration, independent of SPC treatment. If SPC-enhanced actin dynamics were the predominant reason for the observed increase in migration speed on 2D surfaces, this should also lead to an increased speed inside the channels which was not observed. Thus, SPC facilitates the initial invasion into the channel but seems not to affect the further migration speed inside the 3D environment.

Recent research has identified different migration phenotypes in 2D and 3D [32, 61, 107]. In particular, an adhesion independent migration mechanism has been reported for leukocytes in a 3D matrix [61]. Based on a theoretical model such a migration mechanism may be attributed only to partial pressure differences between the leading and rear edge without the necessity of specific cell-surface adhesion [86]. With the migration chip this particular migration mechanism was tested and two distinct migration phenotypes were revealed that are proposed to occur in either a 2D or 3D environment. Doyle *et al.* observed a coordinated migration in fibroblasts moving on thin 1D lines with a width of up to 5 μm [32]. They argue that this behavior depends exclusively on the width of the adhesive lines and resembles the behavior in 3D. Similarly to their experiments, it was found in this work that the majority of the cells showed a 2D-typical *push-and-pull*-like movement when migrating on 7 μm wide lines. However, inside the channels with the same width the majority of the cells showed a *sliding*-like movement. Therefore, it is proposed that the transition from a 2D to a 3D or 1D migration mechanism depends not only on the line width but is also determined by the contact area with the environment, this case the channel walls.

In summary, a detailed investigation was made to study migration dynamics of human pancreatic cancer cells inside microchannels with a particular focus on the effects of keratin reorganization induced by the bioactive phospholipid SPC. Beyond previous knowledge, it was demonstrated that the SPC treatment of Panc-1 cells increases their ability to *invade* and *permeate* narrow channels. Hence, the study may contribute to a more detailed understanding of how cancer cells invade the surrounding tissue and also escape from a primary tumor (*permeate* through the stroma), the first step in tumor spreading. Furthermore, the existence of two different migration phenotypes depending on the dimensionality of the cell environments is shown. Thus, the migration chip provides an easy-to-use experimental tool to promote current research on 3D migration behavior on a single-cell level.

7.2 Nanopatterned microchannels

Micro-sized grooves were etched into silicon substrates and patterned with of gold nanoparticles via block copolymer micellar nanolithography (BCMN). These nano-patterned microstructures can be directly used as a mold for casting microstructured PEG hydrogels while simultaneously transferring the gold patterns to the surface of the hydrogel within one polymerization step. The PEG hydrogel is repellant to protein and cells adhesion while the gold particles can be easily functionalized by covalently binding small molecules or protein fragments. Thus, the presented combination of anisotropically etching silicon substrates with BCMN is a further development, extending the possible applications of the BCMN [53] towards nano-patterned 3D microstructures. However, BCMN on 3D microstructures did not lead to hexagonally ordered patterns of gold nanoparticles. The evaporation of the solvent is the crucial step for distributing the metal salt containing micells homogeneously on the surfaces. On non-flat surfaces this evaporation is locally influenced by the surface topography as capillary forces along corners are reducing the evaporation speed which leads to a local increase in micelle density and multi-dot formation during the plasma treatment.

Part III

Collective cell migration on photo-switchable substrates

Chapter
8

Motivation and experimental approach

Many cell biological phenomena that are observed in *in vitro* studies are influenced by the spatial organization of the cells, the interactions of the cells with artificial surfaces and the interactions among neighbouring cells.

It has been shown for fundamental processes that cell adhesion restricted to geometrical confinements, provided by patterns of adhesion-mediating ligands on biologically inert surfaces, can significantly influence cell behavior. On a single-cell level, for instance, the size and shape of adhesive areas provided to endothelial cells was shown to govern whether the cells would grow or die [45]. Another example are multicellular ensembles that grow in a confined geometry: In endothelial and epithelial cell sheets that were grown under geometrically confined conditions, it was reported that regions of highest accumulation of proliferation events were found to be located along the edges of the cell sheets where the cells experience the highest tensional stress [48].

The cell-adhesive patterns for these experiments were obtained by μCP, a method that allows for patterning surfaces with patterns of adhesion-mediating molecules at a resolution down to a few micrometers. However, a main drawback of patterning surfaces via conventional μCP is, the fact that as soon as cells are seeded on the patterned surfaces, no change in the size of the adhesive areas is possible. This limitation of the μCP technique is excluding its application for the study of migration and expansion of cells that are only initially confined to specific areas.

However, different attempts have been made in the past to dynamically convert a non-adhesive surface into an adhesive one under cell culture conditions in order to study cell migration from spatially confined starting conditions. Following the basic principle of the scratch assay, as described in section 1.2.1, a confluent layer can be, for example, manipulated by disrupting and removing cells via a microfluidic sheer flow [108] or electroporation of the cell membranes in selected regions [109]. Other, less disruptive, methods are based on the principle that cells are initially seeded in restricted areas where cell migration starts upon removing either a teflon barrier [110], a PDMS stencil [111–113], by opening a microfluidic channel [114]. Alternatively, electrochemically induced desorption of entire passivated areas [115, 116] or pre-defined regions [117] were used in combination with μCP techniques to study cell migration dynamics from confined starting conditions. However, the usage of these methods is still restricted either by their limited flexibility choice of shapes for the restricted areas, or by the fact that release of the confinements still disrupts some of the cells, especially those located along the boundaries of the confined areas.

Therefore an improved assay to study collective cell migration in particular or other cell biological phenomena in general, that can be investigated on micropatterned adhesive substrates, has to meet with the following prerequisites:

- Flexible choice of pattern geometries,

- Ability to pattern large areas in a short time,

- High stability of the passivated areas,

- Dynamical and locally resolved switching of non-adhesive into adhesive areas,

- Non-invasive switching process (without any mechanical interference),

- Incorporation of the setup into a microscope setup to directly monitor cell behavior.

To comply with these requirements, a novel experimental setup was developed that enabled to study collective epithelial cell migration from precisely confined cell sheets using a photo-switchable surface passivation technique. The key feature of this setup is a passivating PEG monolayer that is covalently bound to glass substrates via a photo-cleavable group and sensitive to UV light [118, 119].

Many cell types such as epithelial cells within a confluent layer are highly interconnected and exhibit individual and coordinated cell motility phenomena. This includes, for example, membrane protrusions of single cells, as well as synchronized movements of many cells over distances of several cell diameters [81, 113, 120, 121]. In experiments studying collective cell migration, the edges of epithelial cell sheets expand typically in a non-linear way [113, 122]. Epithelial cells at the boundaries are, thereby, able to separate into fast moving cells with active lamellipodia protrusions forming tips of boundary outgrowths ("leader cells") and cells along the sides and behind of these outgrowths, showing little to no lamellipodia activity ("follower cells") [22, 113, 123, 124]. However, it remains poorly understood which cell-internal and external factors influence this separation into leader and follower cells. In particular it is not known, how the geometry of the cell sheet affects the appearance of leader and follower cells. To tackle this question, highly controlled starting conditions regarding the geometry and size of the cell clusters are needed at the point where collective migration is initiated.

The high flexibility of pattern design and the accuracy of the developed *in vitro* migration assay was used to study collective cell migration of MDCK epithelial cells, after being released from initially circular adhesive patterns of different size. In particular, it was investigated how cell collectives of different size expand into a freely available surface area. Accordingly, it was studied how geometry, especially curvature, of cluster boundaries influences and determines the cells' fate and the appearance of leader cells and dynamics of cell sheet expansion was quantified. Furthermore, it was analyzed in which way the incubation time of the cells in their spatial confinement alters cell-cell interactions and modulates their collective behavior. By controlling and varying the geometry of the initial adhesive area it was possible to directly relate characteristics of cell behavior along cluster boundaries to the initial incubation time, size, and shape of the cell sheet.

Chapter 9

Materials and methods

9.1 Preparation of photo-switchable substrates

The single steps of the surface functionalization to obtain photo-switchable substrates is depicted in Figure 9.1. First, glass coverslips with a thickness

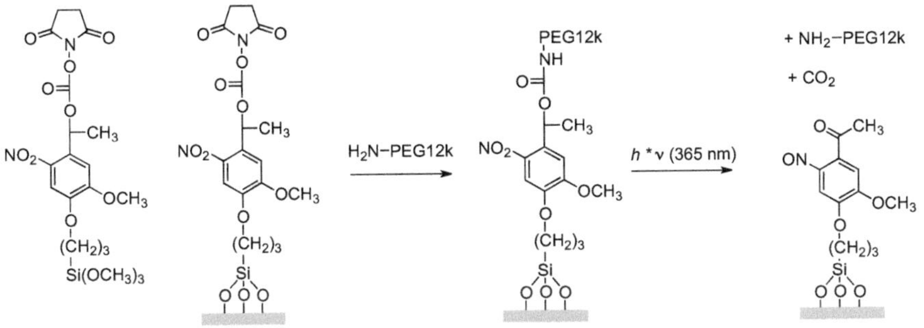

Figure 9.1: Schematic of the surface functionalization reactions. The photo-cleavable silane coupling agent (**5**) is first bound to an activated glass surface. After reaction with a primary amino-PEG the surface becomes biopassivated. Irradiation with UV-light leads to a photo-cleavage of the 2-nitrobenzyl in which the PEG is detached from the surface.

of 0.12–0.17 mm (Matsunami, Japan) were cleaned and activated (hydroxyl-terminated) with piranha solution (sulfuric acid/hydrogen peroxide, 7:3) for 1 hour at 100°C; all chemicals used, if not indicated differently, were purchased either from Wako (Japan), TCI (Japan) or Sigma-Aldrich (USA). After being cooled to room temperature, the substrates were rinsed with deionized water and dried under nitrogen flow. The photo-cleavable linker 1-[5-methoxy-2-nitro-4-(3-trimethoxysilylpropyloxy)phenyl]ethyl N-succinimidyl carbonate (**1**) which was provided by our collaboration partner Professor Kazuo Yamaguchi at the Kanagawa University, Japan (31 mg, 0.3 mmol; detailed synthesis protocol is provided in Appendix C and [118,119]) was dissolved in dry toluene (200 mL) with 0.01% (v/v) acetic acid (20 μL) and the freshly cleaned glass substrates were immersed into this solution for 1 hour at r.t. under nitrogen atmosphere. The silanized substrates (**2**) were rinsed tree times with toluene and three times with acetonitril, sonicated in acetonitril for 15 min and dried under nitrogen flow. In order to make the surface non-adhesive for cells, PEG12k-amine was bound to the photo-cleavable linker on the glass surface by immersing the silanized glass substrates in 200 mL acetonitril containing triethylamine (1.43 μL, 0.05 mmol) and PEG12k-amine (120 mg, 0.05 mmol, Sunbright MEPA-12T; NOF Corporation, JAPAN) over night at r.t. under nitrogen atmosphere. Finally, the substrates (**3**) were washed with acetonitril, sonicated for 10 min, dry blown with nitrogen and stored protected from light under nitrogen atmosphere until further experiments were made. Irradiation of the surface with UV-light triggers the cleavage of the linker (**4**) while releasing the ammino-PEG. The remaining surface (**4**) shows an increased hydrophobicity compared to the PEG-terminated one (change in contact angle from 47° to 60°) [118] and becomes cell adhesion mediating as depicted in Figure 9.2.

9.2 Experimental setup

Figure 9.3 shows the stepwise UV irradiation and cell seeding procedure that was used to study cell behavior on micropatterned surfaces. First, the homogeneously covered and functionalized glass coverslip with the PEG covalently bound to the surface via the photo-cleavable linker were cut into small pieces of approximately 5x5 mm which were immersed for sterilization in 70% ethanol for 5 min, dried

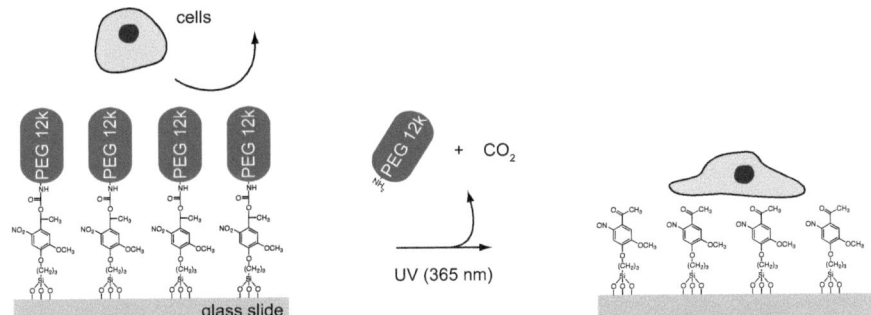

Figure 9.2: Illustartion of the working prinicple of photo-switchable surface passivation. Cell adhesion on the surface is initially inhibited by a monolayer of PEG that is covalently bound to the surfave via a photo-cleavable linker. Cleavage of the linker that is sensitive to UV light releases the PEG and the surface becomes cell adhesion mediating at the irradiated regions on.

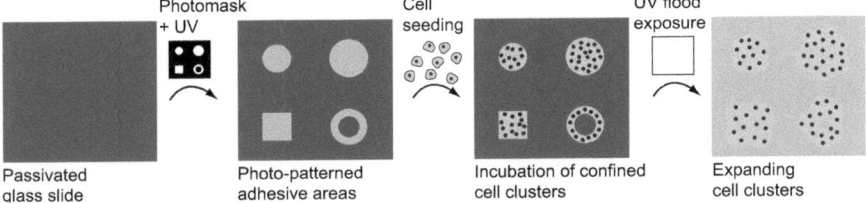

Figure 9.3: Stepwise UV irradiation and cell seeding procedure for experiments with photo-switchable substrates. All steps can be carried out and monitored with the substrate mounted on an inverted microscope.

under air flow and glued with Twinsil (Picodent, Germany) on a glass-bottom dish (MaTek, USA). The glass-bottom dishes were filled with PBS and placed on the stage of an inverted microscope (XI81, UPlanSApo 10x/0.4; Olympus, Japan or AxioObserver, Fluar 10x/0.5; Carl Zeiss, Germany), equipped with a mercury arc lamp (HBO 103W/2; Osram, Germany). By inserting a photomask into the position of the field diaphragm of the reflected light, the patterns on the photomask were projected onto the glass surface using a conventional filter cube, normally used for fluorsecent imaging (excitation G365, beam splitter FT395, emission BP445/50, 49HE; Carl Zeiss, Germany) as depicted in Figure 9.4. Irradiation of the surface

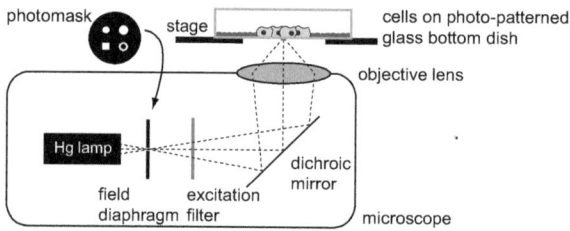

Figure 9.4: Illustration of the image-projection exposure setup implemented into an inverted microscope. Patterns on the photomask are focused on the substrate through the objective by inserting the mask into the light path at the position of the field diaphragm. Figure adapted from [125].

with UV-light dose of 10 J/cm^2 triggered the cleavage of the the PEG-chains, thus making the surface cell adhesive. After replacing the PBS with normal cell culture medium, cells were seeded onto the photo-patterned coverslips and experiments were carried out as described in the section below.

Prior to each exposure cycle the UV-intensity at the focal plane was measured with a UV-meter (UIT-250 UVD-S365, $\lambda = 300\text{-}400$ nm; Ushio, Japan or OMM-610B with a OMH-6732B Silicon PowerHead 350-530 nm; ILX Lightwave, Germany). The photomasks used in these experiments were high resolution (5080 dpi) printed transparencies commonly used for offset printing (Heidelberg Linotype Signasetter Pro on high-end Fuji-film; Printfactory GmbH, Germany). The projection exposure with the samples mounted inside the microscope allowed for fast and multiple patterning of several patterns at different positions.

9.3 Cell experiments

9.3.1 Cell culture

MDCK epithelial cells and MCF-7 breast cancer cells were obtained from RIKEN cell bank, Japan. Cells were cultured at 37°C, 5% CO_2 in minimum essential medium (MEM) (#M4655; Sigma, Japan) supplemented with 10% heat inactivated FBS (#S1820-5009; BioWest, Japan), 1% MEM non-essential amino acids (#11140; Sigma, Japan), 1% penicillin-streptomycin (#168-23191; Wako, Japan) and freshly added L-glutamin (#G7513; Sigma, Japan). The MCF-7 medium was additionally supplemented with 1% sodium pyruvate (#190-14881; Wako, Japan). For subculturing and prior to experiments, cells were detached with 0.25% trypsin-EDTA (#T4049; Sigma, Japan).

9.3.2 Cell patterning and photo-induced migration

Cells were seeded on the photo-patterned substrates at a density of $3.5*10^5$ cells/cm^2. One hour after seeding, non-attached cells were removed by changing the medium. After an incubation time, typically between 9 and 38 h, a second UV-light exposure with a dose of 10 J/cm^2, this time of the whole substrate (flood exposure), released the confinement and triggered cell migration. Irradiating the cells a single time with 10 J/cm^2 is not harmful and showed no mayor effect on cell migration behaviour, as observed during the experiments and quantified previously [126].

Phase-contrast live-cell imaging was performed in a heated and air-humidified chamber (INU-ONI-F1, Tokai Hit, Japan) on the motorized stage (Molecular Devices, USA) of an automated inverted microscope (XI81, UPlanFLN 10x/0.3 Ph1; Olympus, Japan) and controlled with the MetaMorph software (Version 7.6.3.0; Molecular Devices, USA) while images were captured with a CCD camera (Retiga EXi; QImaging, Canada). The motorized stage enabled the observation of multiple positions in one experiment. For time-lapse videos, images were taken every 2–4 minutes at the selected positions over a period of 2–24 hours.

9.3.3 Immunocytochemistry

Filamentous actin, E-cadherins and nuclei of fixed MDCK cells were fluorescently labeled and imaged under the microscope; for the detailed steps of the staining protocol see also Table 9.1. Cells were fixed with 4% PFA (#P6148; Sigma, Germany) in PBS and permeabilized with 0.1% Triton-X100 (#T8787; Sigma, Germany) in PBS and blocked with bovine serum albumin (BSA) (#A2153; Sigma, Germany). E-cadherin localization was detected with the monoclonal DECMA-1 antibody (#ab11512; Abcam, UK) in combination with Alexa488 donkey anti rat (#A21208; Life Technologies, Germany) antibody; filamentous actin was visualized with AlexaFluor568 phalloidin (#A12380; Life Technologies, Germany)and nuclei were stained with 4',6-diamidino-2-phenylindole dilactate (DAPI) (#D3571; Life Technologies, Germany). Images were acquired with an AxioCam MRm3 CCD-camera on an AxioExaminer upright microscope equipped with a W-Plan-Apochromat 40x/1.0 objective and a Colibri LED illumination system (all Carl Zeiss, Germany).

9.3.4 Labelling cells for fluorescence live-cell imaging

The cytosol and nuclei of MCF-7 were fluorescently labeled for live-cell tracking. A stock solution of CellTracker Green CMFDA (#C7025, Life Technologies, Japan) was obtained by dissolving 50 μg CMFDA in 10 μL DMSO. In order to obtain a 5 μM working solution, 1 μL of the stock solution was added to 2 mL serum free culture medium. Cells were incubated with this staining solution for 15 min at 37°C; subsequently the medium was replaced with normal culture medium and after an incubation time of 30 min, cells were washed with PBS once and phase-contrast live-cell imaging was started with standard cell culture medium. Directly before image acquisition started, Hoechst 33342 (#H3570; Life Technologies, Japan) with a final concentration of 0.2 μg/mL was added to the medium.

Table 9.1: Staining protocol for MDCK cells. Labeling of E-cadherin, filamentous actin and nuclei.

wash	1x		37°C	PBS
fix	1x	15 min	4°C	4% PFA in PBS
wash	3x	2 min	r.t.	PBS
permeabilize	1x	exact 3 min	r.t.	0.1% Triton X-100 inPBS
wash	3x	2 min	r.t.	PBS
block	1x	30 min	37°C	3% BSA in PBS
wash	1x	2 min	r.t.	PBS
first antibody	1x	over night	4°C	rat α-E-cadherin DECMA-1 1:1000 in 1% BSA in PBS
wash	3x	15 min	r.t.	PBS
second antibody	1x	60 min	37°C	AlexaFluor 488 donkey α-rat IgG 1:500; AlexaFluor 568 phallodin 1:40; DAPI 1:500 in 1% BSA in PBS
wash	3x	15 min	r.t.	PBS

9.4 Image analysis and data processing

9.4.1 Automated image segmentation

Phase-contrast images of the cell clusters were binarized using automated routines written in MATLAB (Version 7.5; The MathWorks, USA). First, a sobel filter was applied to detect intensity gradients in the image and it returned the edges where the gradients had their maxima as a binary image. MATLAB automatically calculates a certain threshold gradient above which edges are identified. This threshold was fine-tuned for every individual time-lapse video. The edges, obtained from the sobel filter were modified by subsequent steps of dilating, closing, thinning, and eroding as provided by the MATLAB code in the Appendix D.1).

In the dilation process, a linear structuring element with a length of 5 pixels was used to probe and expand the shapes in the input image.

With the closing routine the space between all interconnected regions was filled, which was suitable for circular cell sheets, but it would also fill the inner hole of donut-shaped patterns. For donuts, a different strategy was chosen where the hole of the donut was automatically detected in the first frame of the image sequence and the holes of the following frames were identified by choosing the object that had the highest overlap with the hole from the previous frame. The thinning of the detected object was applied to preserve thin boundary shapes that were masked in the dilation step.

For the final eroding step, a diamond-shaped structuring element with a diameter of 2 pixels was used to probe and smoothen the shapes in the binarized image.

The obtained binary image shows the size and shape of the cell sheets. Each automatically analyzed image was interactively examined and, if necessary, manually corrected.

9.4.2 Cell densities

The initial mean cell areas within each cell sheet were calculated in order to assure comparable starting conditions among the evaluated cell sheets. Only clusters with a initial mean cell areas in the range between 440 and 900 μm^2/cell were selected for the later analysis. To obtain the values of initial mean cell area for each cell cluster, the surface area of the whole cell cluster, obtained from the automated image segmentation, at the time point of confinement release was divided by the initial number of cells that was manually counted in the phase-contrast images.

9.4.3 Spatiotemporally resolved migration dynamics

For angular resolved protrusion activities the center of mass of the cell cluster in the first frame of the segmented time-lapse video was calculated. Its position was used as origin for the following analysis. Protrusion rates of the membrane, towards or away from the origin, were calculated along angular segments with a resolution

IMAGE ANALYSIS AND DATA PROCESSING   83

of 1°, starting at the top and turning in clock-wise direction (see MATLAB code provided in the Appendix D.2). Averaged displacement rates of the boundaries were calculated from 4 minutes intervals while images were recorded at two minutes per frame which allowed for reducing the background noise that was mainly caused by the automated image segmentation.

A surface plot, color-coded for the calculated protrusion rates, was plotted to visualize the protrusive activities of each angular segment over time. Positive protrusion rates (warm colors) indicate to a movement away from the center and negative values (cold colors) indicate a movement towards the origin.

9.4.4 Cell sheet expansion

In order to quantify the temporal expansion of the circular cell sheets, mean cluster radii were calculated at each time point. Therefore, the cluster area, obtained from the segmented time-lapse videos, was used to calculate a mean radius using the geometric equation: $r = \sqrt{\frac{A}{\pi}}$, with the radius (r) and the area (A). From this data, the increase in mean cell area was calculated between the time point of confinement release and two hours later.

9.4.5 Leader cell quantification

Leader cells were identified by their active lammellae protrusions at the clusters' boundaries. The protrusion dynamics were manually observed in the time-lapse videos to identify the leader cells one hour after the confinement was released. In order to compare the amount of leader cells that evolve at the cell clusters' boundaries of different initial sizes, the number of leader cells was normalized to a frequency of leader cells per perimeter unit by dividing the number of counted leader cells with the initial perimeter of the cell cluster:

$$\begin{aligned} f_{LC} &= frequency\ of\ leader\ cells\ per\ perimeter\ unit \\ &= \frac{number\ of\ leader\ cells * 100\ \mu m}{initial\ perimeter} \end{aligned} \quad (9.1)$$

Chapter 10

Results

10.1 Cluster expansion of MDCK cells

10.1.1 Photo-controlled cell adhesion and triggered cell migration

In order to control cell adhesion and trigger migration of MDCK cells, glass slides covered with a photo-removable PEG monolayer, were used. The PEG was covalently linked to the glass via a photo-cleavable 2-nitrobenzyl group, sensitive to UV light. This property allowed for conversion of the initially passivated surface into a cell adhesion mediating surface by releasing the amino-PEG into the solution as schematically depicted in Figure 9.2 and published previously [118]: Passivated glass coverslips were patterned with adhesive areas upon irradiation with UV light with a wavelength of 365 nm. Cells seeded on these surfaces adhere and grew only on the initially irradiated areas. A second irradiation step, which was not harmful for the cells that already grew on the surface, converted the still inert, PEG-covered, areas within seconds into adhesive ones with the sheets expanding into the idle areas. Hence, these photo-sensitive surfaces permitted not only an incidental patterning, but also the study of the dynamical behavior of cell clusters that were released from spatially well-defined confinements. The sequential steps were carried out with the substrates mounted on the microscope while simultaneously monitoring the cells via live-cell imaging.

Only the light path of the microscope limited the resolution of the pattern size and the passivation with PEG-12k was very stable, permitting even long-term experiments of several days to weeks under normal cell culture conditions as also previously reported [118].

Examples of cell ensembles confined in the different circular and donut-shaped patterns that were used in the following migration analysis are shown in Figure 10.1. These images show cells that were growing for 20–25 hours on the initially photo-patterned areas where they formed confluent cell sheets. UV flood exposure of these substrates would cleave the PEG from the still passivated areas and initiate cell expansion.

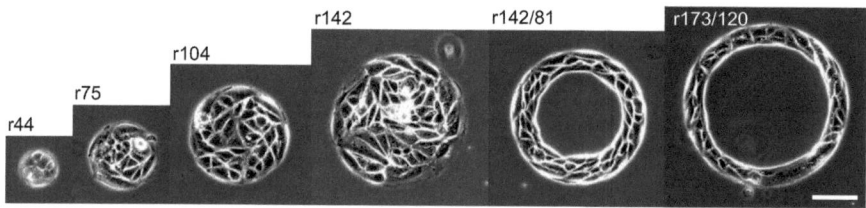

Figure 10.1: Phase-contrast images of MDCK cells growing in confined photo-patterned areas. Radii of the cell sheets are indicated in the upper left corner of each image. Scale bar, 100 μm.

It was also possible to obtain cell clusters that grew in more complex structures like the logo of the Max Planck Society (Minerva's head) as it is shown in Figure 10.2. photomask. The image projection exposure of the photomask onto the substrate is very precise. In this example, the smallest feature size is 14 μm with cells adhering on it. The deviation of the feature size on the photomask from the finally covered cell pattern is ±2 μm. An overview of the geometric parameters of circular and donut-shaped confined cell clusters used in this study are summarized in the Appendix B.1.

Phase-contrast images of different time points in a typical experiment are shown in Figure 10.3 for a donut-shaped cell sheet (taken from Video A.3.1): First, the cell cluster is trapped in a donut-shaped adhesive area and the confinement is released via UV flood exposure of the substrate. Subsequently, the cell sheet reacts to the

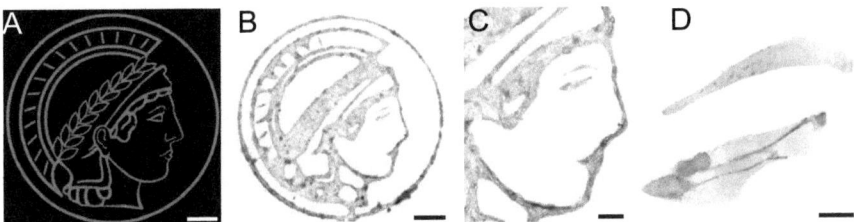

Figure 10.2: Example of a complex microstructured cell pattern. (A) Photomask with the logo of the Max Planck Society, Minerva's head. Scale bar, 1 mm. (B-D) Fluorescence microscopy image of MDCK cells growing on the adhesive surface pattern with the shape of the Minerva's head. Image is shown at different magnification levels. E-Cadherin localization is labeled with antibodies (blue) and nuclei are labled with DAPI (yellow). Scale bar, 250 μm (B), 100 μm (C), 25 μm (D).

released confinement and expands on the newly available surface.

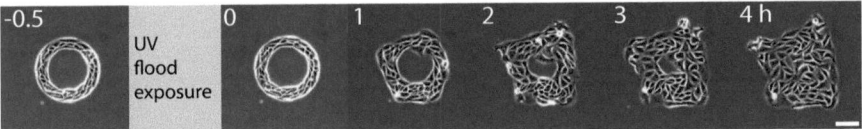

Figure 10.3: Example of cell sheet expansion on photo-switchable surface pattern. Cells are seeded on a photo-patterned surface and adhere within the donut-shaped confinement (-0.5). Upon UV flood exposure the initial confinement is released (0) and cells start to migrate and expand into the readily accessible area (1-4). For the whole time-lapse experiment see Video A.3.1. Numbers indicate the time point in hours before and after flood exposure; scale bar, 100 μm.

An example of triggered cell sheet expansion from a stripe-pattern with almost linear boundaries is shown in Figure 10.4 and Video A.3.2. It can be noticed that 12 hours after confinement release, the MDCK cells are evenly distributed due to their high motility within the cell sheet, leading to a homogeneous cell density over the whole cell sheet. It can be also seen that the finger-like outgrowths preferentially evolve from the regions with highest boundary curvature (the non-linear regions of the cell cluster). MDCK cell clusters expanding from a more complex pattern, the shape of the Max Planck Society logo, are shown in Video A.3.3.

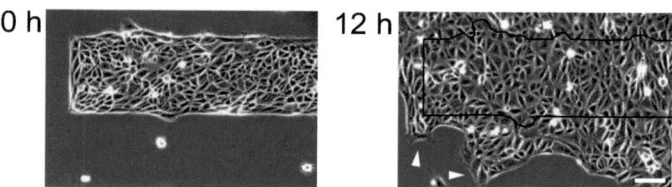

Figure 10.4: Cell sheet expansion from a cluster with linear boundaries. (0 h) Cell sheet expanding from a stripe-shaped cell cluster with initially almost straight boundaries, directly after confinement release. (12 h) After 12 hours of expansion, characteristic finger-like outgrowths appear, indicated by the white arrows. The black line indicates the initial boundary of the cell cluster. Initial width of the stripe was 250 μm; scale bar, 100 μm

In order to assure comparable starting conditions for various experiments the initial cell number and cluster area for all evaluated circular cell clusters was determined. For the later data processing and analysis, only cell sheets with comparable cell densities were selected. Figure 10.5 A shows the initial mean cell area of the cells within the different circular cell sheets, that was selected to be in the range between 440 and 900 μm^2/cell. Accordingly, a linear increase of initial cell number with increasing cluster area was observed for all selected and later analyzed circular cell sheets, compare Figure 10.5 B.

10.1.2 Expansion characteristics of cell clusters

The impact of initial cell cluster size on the migratory behavior was studied by monitoring the expansion of circular cell clusters with initial mean radii of 44, 75, 104, and 142 μm within the first 2 hours after confinement release. Boundary outlines were obtained at each time point (see Figure 10.6 A and C).

In order to visualize the movement dynamics of the boundaries, the protrusion and retraction speeds were calculated at the boundaries in an angular resolved

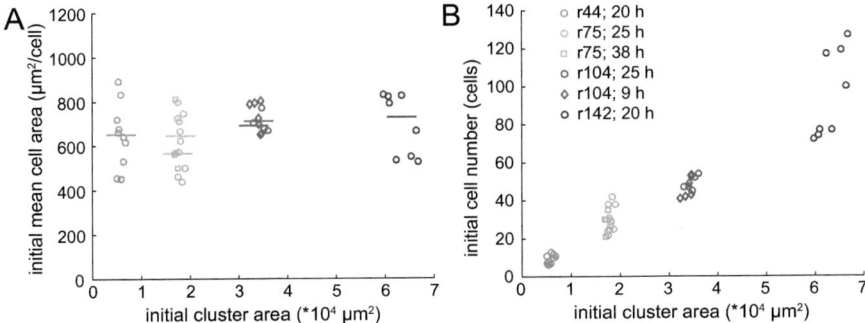

Figure 10.5: Initial cell numbers and cell densities of evaluated cell sheets.
(A) Linear increase of initial cell number and cluster area for all evaluated circular cell sheets. (B) The initial mean cell area was similar for all evaluated circular cell sheets with values between 440 and 900 μm^2/cell. Lines indicate the mean values of each group.

way and plotted versus the time points after confinement release (Figure 10.6 B and D); warm and cold colors represent protrusion and retraction rates, respectively. As it has been reported for linear cluster boundaries [113], the edges of cell sheets expand in a non-homogeneous way with some of the cells migrating faster than others, which leads to higher fluctuations in the protrusion rate plots (Figure 10.6 B and D). Comparing the changes in average cluster radius over time for differently sized circular clusters the following expansion characteristics can be observed (Figure 10.7 A). The mean cluster radius increases over a time period of 2 hours after a characteristic delay time of about 15 minutes independent of the initial cluster size. However, significant differences in cluster expansion are found if the initial incubation time is reduced from 20-25 hours to 9 hours only. In the latter case, the mean cluster radius increases faster and to about a 1.25-fold higher value than in the case of 25 hours incubation for the same initial cluster size (r104, Figure 10.7 A, Video A.3.4). The cell sheet expansion after 9 hours incubation only, starts nearly without delay after confinement release (Figure 10.6 D and 10.7 A). For an extended incubation time of 38 hours, no change in the cluster expansion rate is observed (r75, Figure 10.7 A).

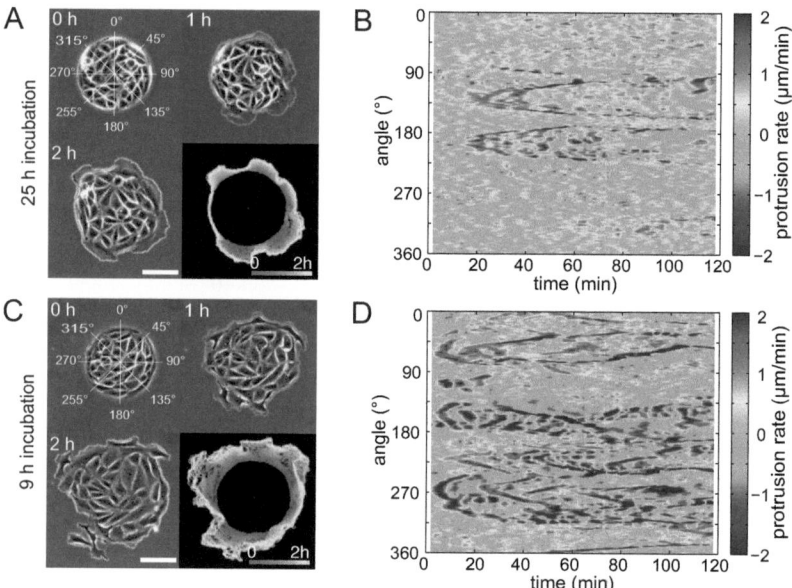

Figure 10.6: Expansion dynamics of circular cell clusters. (A, C) Cell expansion from a circular cell cluster with an initial radius of 104 μm, incubated for 25 hours (A) or 9 hours (C) before confinement release. A color-coded overlay of all cluster boundaries from the first 2 hours after releasing the confinement is inserted in the lower right of each image. The full dynamical behavior can be followed in Video A.3.4. Scale bars, 100 μm. (B, D) Spatiotemporally resolved protrusion activity of the cell sheet boundary. Positive protrusion rates (warm colors) refer to a movement of the boundary away from the center, negative (cold colors) towards the center.

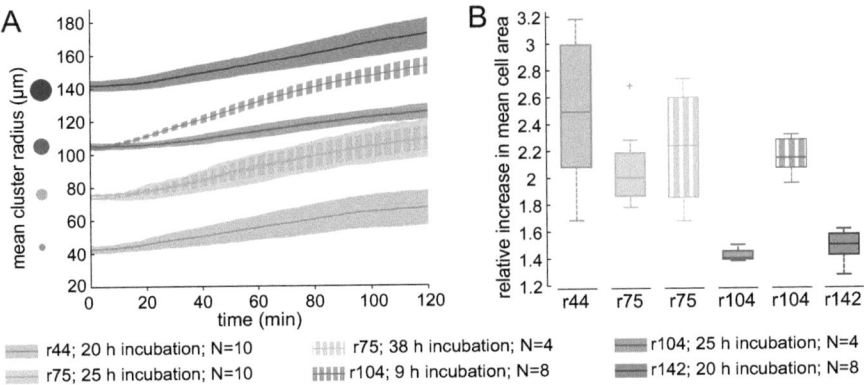

Figure 10.7: Quantification of cell sheet expansion dynamics. (A) Temporal characteristics of the calculated mean radii of the cell cluster after confinement release for different initial sizes and preceding incubation times. Cells were incubated between 20 and 25 hours (thick lines), 9 hours (thin red line) or 38 hours (thin yellow line). Mean values of 4-10 clusters per cluster size are given with errors as s.d.s. (B) Increase in mean cell area within 2 hours after confinement release for cell clusters of different circular size and incubation times. Statistics: Boxplot is representing the four quartiles of each data set, with the box comprising the two middle quartiles, separated by the median. The upper and lower quartiles are represented by the single lines extending from the box and their length is restricted to 1.5 times the length of the box. Outliners are indicated as individual points.

In order to analyze if all cells, in average, contribute equally to the increase in cluster size, the increase of the mean cell area was determined after 2 hours (Figure 10.7 B). Starting with an initial mean cell area of 440 to 900 μm^2/cell, comparable for all cell clusters (Figure 10.5), this value increases in the process of cell sheet expansion as hardly any cell division occurs within 2 hours of observation. However, two characteristic regimes were observed depending on the initial cluster size: For small initial cluster sizes (r44, r75) the mean cell area increases by a factor of 2.5 and 2.0, respectively, while for larger cluster sizes this increase is less drastic (1.4 fold for r104, and 1.5 fold for r142, Figure 10.7 B). However, at an incubation time of 9 hours only, the mean cell area increases by a factor of 2.1 even for the bigger cell cluster (r104). This result suggests that in small clusters all cells contribute to the expansion since they are close to the boundary. For larger clusters, mainly cells in proximity to the boundary increase in area whereas they expand to a smaller degree in the interior of the cluster, if incubated for at least 20 hours (Video A.3.4). These results show, that cell sheet expansion is directly dependent on initial cluster geometry and incubation time as it is supported by the further findings.

10.1.3 Boundary curvature affects leader cell formation

As the cell sheet expansion was monitored immediately after confinement release, it was possible to observe the lamellipodia activity and categorize the cells into leader cells (expanding, highly active lamellar protrusions) and non-leader cells (no lamellar protrusions). In order to investigate how much the initial cell cluster size affects the frequency of leader cell appearance the amount of leader cells was quantified one hour after confinement release. The absolute median values for the occurrence of leader cells are given in the Appendix B.2. To compare leader cell appearance between different cell sheet sizes, the appearance of leader cells was normalized as defined in section 9.4.5.

The results for cells incubated on the photo-patterned surfaces for 20-25 hours prior to the release of the confinement are plotted in Figure 10.8. No significant difference in f_{LC} is observed for the two small cell clusters of r44 and r75 (median

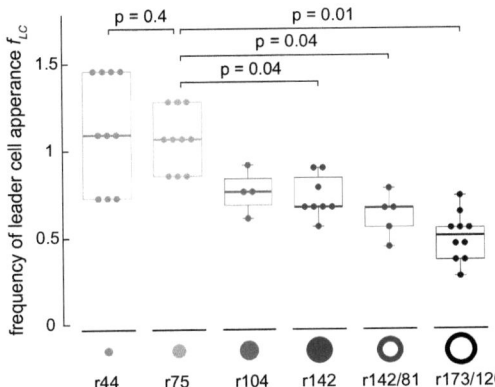

Figure 10.8: Frequency of leader cell appearance at circular and donut-shaped patterns of different sizes. Frequencies of leader cell appearance per perimeter unit f_{LC} were obtained for cell sheets of different size and shape. To obtain f_{LC} the number of leader cells at the cell sheet boundaries was counted one hour after confinement release and divided by the initial perimeter of the cell sheet, as explained in section 9.4.5. Statistics: The graph shows an overlay of the single frequencies obtained (dots) with a boxplot indicating the median values; Wilcoxon-Mann-Whitney-Test with p-values as indicated.

f_{LC} of 1.09 and 1.07, respectively). If the initial radius of the cell cluster increases to 104 μm and 142 μm, however, a decrease in median f_{LC} was observed from 1.07 (r74) to 0.76 (r104) and 0.67 (r142). By increasing the radius of the cell sheet, the area and thereby the total number of cells in the cluster increases while the curvature of the boundary decreases. To test whether it is the increasing number of cells that can interact as a collective, or only the boundary curvature of the cell sheet that influences the separation into leader and follower cells, the total number of cells in the sheet was artificially decreased by using donut-like ring patterns. Keeping the outer radius constant with r = 142 μm, while reducing the total cell number due to a passivated interior, the frequency of leader cells per perimeter unit remained unchanged with f_{LC} = 0.67 (r142/81) compared to the filled circle (r142). By further increasing the radii of donut structures to r = 173 μm (r173/120), there waas even a further drop in median f_{LC} to 0.55, meaning that less leader cells were formed per perimeter unit.

10.1.4 Cell behavior differs along convex or concave boundaries

In order to understand if curvature orientation of cell cluster boundaries has an impact on leader cell formation, cell behaviors were compared at the inner concave boundary with the ones at the outer convex boundary. In Figure 10.9 A and Video A.3.6 a representative example of a cell sheet expanding from a donut-shaped pattern is shown.

The evolution of the boundaries at the inside and outside of the cell sheet is illustrated in Figure 10.9 B. Particularly striking is the observation that cells at the outside boundary show a separation into leader and follower cells while the cells along the inner boundary are moving towards the center in a rather homogeneous way without leader cells standing out. There are no characteristic regions of altered protrusion rates for the inside case in Figure 10.9 B, a behavior comparable with observations of a purse-string mechanism in wound closure [127–130]. Differences in boundary cell behavior between convex and concave cluster boundaries were even observed before the confinement of the cell cluster was released. Figure 10.10

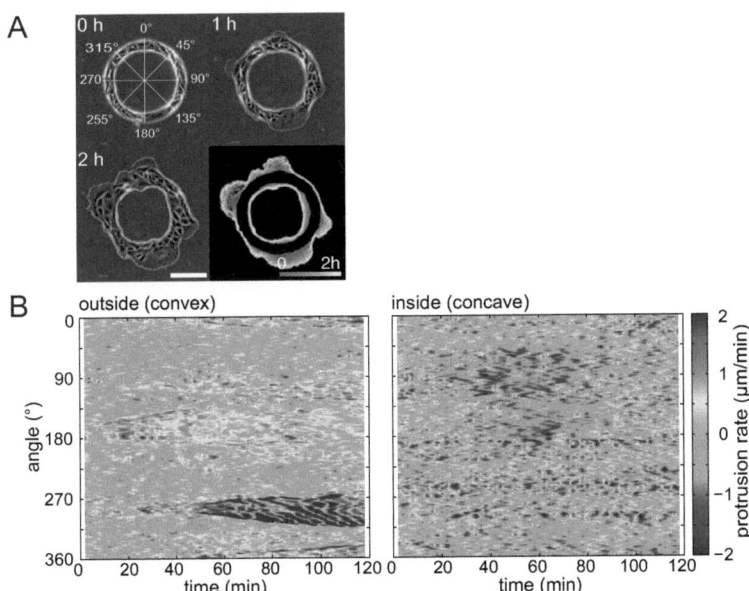

Figure 10.9: Expansion dynamics of donut-shaped cell clusters. (A) Phase contrast images depicting the expansion of a donut-shaped cell sheet with an initial outer and inner radius of 173 μm and 120 μm, respectively, incubated for 25 hours prior to confinement release. A color-coded overlay of all inner and outer cluster boundaries of the first 2 hours after releasing the confinement is inserted at the lower right of the image (see also Movie S4). Scale bar, 100 μm. (B) Spatiotemporally resolved protrusion activity along the outer convex and inner concave boundaries. Positive protrusion rates (warm colors) refer to a movement of the boundary away from the center, negative values (cold colors) towards the center.

A and B show a still image from Video A.3.5 where a cell cluster is incubated in a donut-shaped geometry over a time period of 5 h. Almost all of the cells along

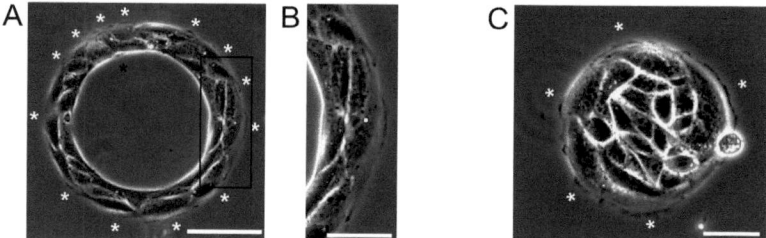

Figure 10.10: Cell behavior is not symmetric for cells along convex and concave boundaries. (A) Confined donut-shaped cell cluster (r142/81); first frame from Video A.3.5. Cells with lamellar protrusions are indicated with white asterisks along the outer boundary and a black asterisk along the pattern inside. Scale bar, 100 μm. (B) Close-up of (A). Scale bar, 50 μm. (C) Confined circular cell cluster (r75); first frame from Video A.3.7. Cells with lamellar protrusions along the convex boundary are indicated with white asterisks. Scale bar, 50 μm

the outer boundary show lamellar protrusions trespassing the boundary, trying to adhere on the inert surface (indicated in Figure 10.10 A with white asterisks), while this behavior is hardly seen at the inner, concave, boundaries (indicated in Figure 10.10 A with a black asterisk).

10.1.5 Incubation time is important for curvature sensing and collective behavior

We directly compared the frequency in leader cell formation along the outside (convex) and inside (concave) boundaries of donut-shaped cell clusters. Frequencies were quantified for cell clusters that had been incubated for different periods of time (9, 25, 38 h) prior to the confinement release. Medians of the absolute number of leader cell per cluster are given in the Appendix B.3.

Frequencies of leader cells per perimeter unit f_{LC}, measured at the inside and outside of the donut-shaped clusters (r142/81) are summarized in Figure 10.11. If

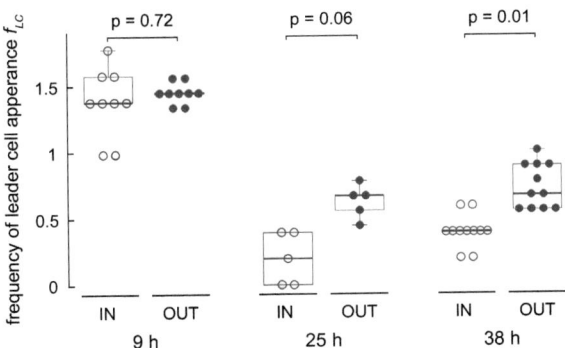

Figure 10.11: Leader cell formation is altered by incubation time and curvature. Quantification of leader cell appearance 1 hour after confinement release at donut-shaped cell sheets with an outer and inner radius of 142 μm and 81 μm, respectively (r142/81). Frequency of leader cell appearance per perimeter unit one hour after confinement release along the inside and outside of the donut was quantified for clusters with three different preceding incubation times (9 h, 25 h, and 38 h). Statistics, boxplot with median, Wilcoxon-Mann-Whitney-Test with p-values as indicated.

cells were incubated for only 9 h, almost all cells along the inner and outer boundaries showed high protrusion frequencies, thus leading to values for the median f_{LC} of 1.38 and 1.45 at the inside and outside, respectively. For incubation times of 25 and 38 hours frequencies of leader cell appearance were more than halved to median values of less than 0.67.

In stark contrast to the behavior after 9 hours incubation, symmetry in leader cell formation at the inside and outside is broken and less leader cells occur along the concave boundary than on the convexly-shaped outside of the cell cluster. For the experiments with 25 hours incubation time, the median f_{LC} is 0.20 for the inside, and 0.67 for the outside, and they are similar after 38 hours with 0.39 and 0.67, respectively. This effect of incubation time on migratory cell sheet expansion is also consistent with the differences in cell sheet expansion of whole circles (Figure 10.6 B, D, 10.7 A, B, and Video A.3.4). To elucidate these differences of cell behavior with varying incubation times, the localization of cell-cell contact forming E-cadherins was visualized via immunofluorescence stainings after incubation times of 9 and 28 hours (Figure 10.12).

In a confluent epithelial-like sheet of MDCK cells, E-cadherins are typically localized along the cell-cell contacts as it is seen in the case after an incubation time of 28 hours, regardless whether the geometric confinement was released or not. However, after an incubation time of 9 hours only, diffuse signals of E-cadherins in the cytoplasm were detected, showing no characteristic localization along the cell-cell boundaries. This observation indicates that the cells had not sufficient time to form and stabilize cell-cell connecting tight junctions via their E-cadherins.

Actin localization was comparable for cell clusters that were incubated for 9 or 28 hours. The signal of the actin localization was characterized by increased, ring-like, intensities (transverse arcs) along the inner and outer boundaries with partly very weak signals, outside these arcs along the outer boundaries of the donuts (Figure 10.12). Once, the confinement was released and cells migrated into the open space for two hours, a break in these arc structures was observed at the tip of the leader cells which was consistent with observations from other epithelial cells [22].

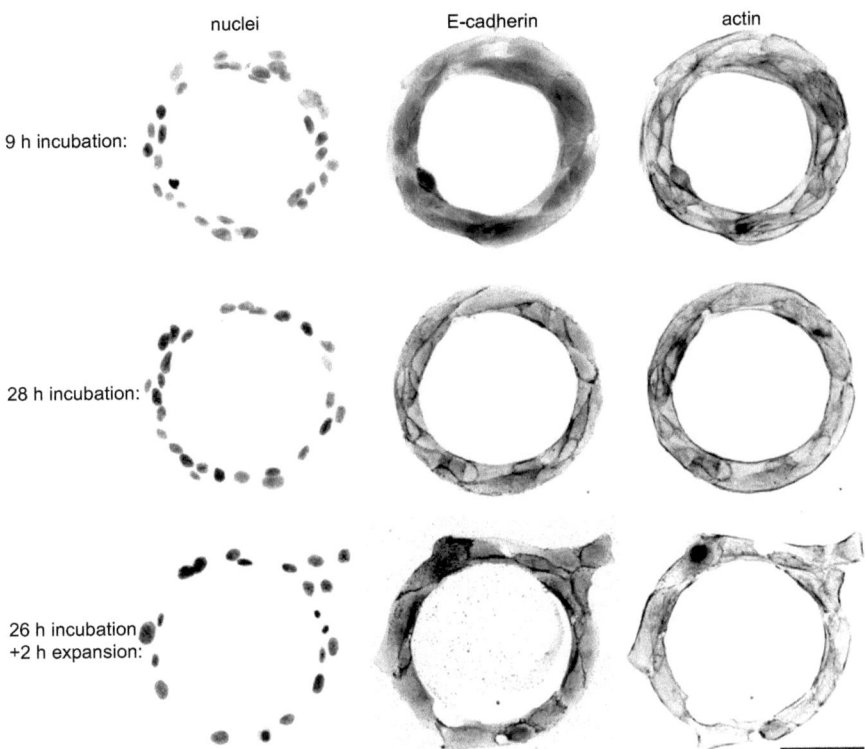

Figure 10.12: **Immunofluorescence images of MDCK cells seeded on photo-patterned surfaces.** Cell sheets were incubated for 9, 28 hours or 28 hours with released confinement during the last 2 hours. Nuclei, E-cadherins and actin stainings were labeled with DAPI, antibodies, and pahalloidin, respectively. Scale bar, 100 μm.

10.2 Expansion behavior of MCF-7 cell clusters

Photo-patterned glass coverslips were used to grow clusters of human breast cancer MCF-7 cells of different shape and size as shown in Figure 10.13 in order to study their expansion dynamics after confinement release. Although MCF-7 cells

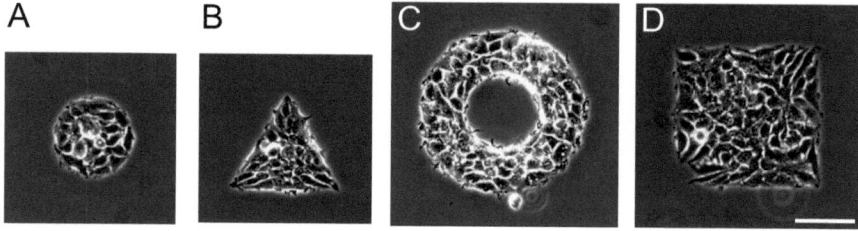

Figure 10.13: Examples of MCF-7 cells seeded on photo-patternened substrates. MCF-7 cell sheets incubated on photo-patterned substrates of the following shapes: (A) Circle with radius r = 70 μm. (B) Equilateral triangle with side length a = 180 μm. (C) Donut-shaped cell sheet with inner and outer radius r = 140/40 μm. (D) Square with side length a = 240 μm. Scale bar, 100 μm.

are known to grow with an epithelial-like morphology forming a monolayer of interconnected cells, they hardly migrate and also the morphology of the cluster boundary is very different than the one of MDCK cells: The boundary MCF-7 cell clusters is very rough with the cells forming only narrow lamellipodia protrusions of a short life-time.

The expansion of cell sheets is mainly driven by cell proliferation within the cluster and cells do not exhibit a crawling-like migration behavior on the surfaces. For example, Figure 10.14 shows a cell sheet that is expanding rather slowly and although the edges of the cell sheet expand in a non-homogeneous way hardly any separation of cells into leader and follower cells is observed, compare also Video A.4.1. It can be further noticed, that 12 hours after confinement release the cell density in the newly occupied area is lower than in the lower part of the image where the cells grew from the beginning. This difference in cell density is contributed to the low motility of the MCF-7 cells and in stark contrast to the motile behavior of MDCK cells which show 12 hours after confinement release a homogeneous cell density over the whole cell sheet (compare Figure 10.4). Further examples of MCF-7 cell sheet expansion from different initial patterned shapes are shown in Videos A.4.2 (triangles), A.4.3 (donut-shape), and A.4.4 (stripe).

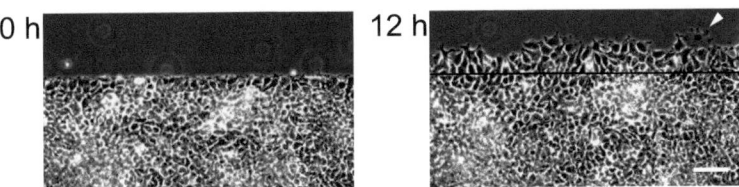

Figure 10.14: MCF-7 cells expanding into free surface starting from a linear boundary. Figure shows two still frames from Video A.4.1 at time point t = 0 and t = 12 hours after releasing the photo-patterned confinement. A rare example of a cell with higher lamellipodia activity at the edge of the cluster is indicated with the white arrow. Scale bar, 100 μm.

MCF-7 cells were used to test, whether it is possible to label the cells with fluorescent dyes in order to follow and analyze the collective migration from the photo-patterned cell clusters. The cytoplasm was stained with CellTracker Green CMFDA which is permeating the cell membrane, incorporated in the cytoplasm, showing fluorescent activity if excited with a wavelength of $\lambda = 492$ nm. Nuclei were stained with Hoechst 33342. Figure 10.15 shows a still frame of the Video A.4.5 where the triangular confinement of the cell cluster was released and cells were free to expand into the free surface while being monitored with phase contrast and fluorescent light. Images were recorded at a frame rate of 5 min/frame (phase-contrast), 15 min/frame (CMFDA) and 30 min/frame (Hoechst). Within the 12 hours of the experiment, the cell cluster hardly expanded and no cell division was observed. Thus, the labeling and exposure of the cells with fluorescent light disturbs the cell functions and the cells' metabolism in a severe way, leading to inhibition of cell proliferation and migration.

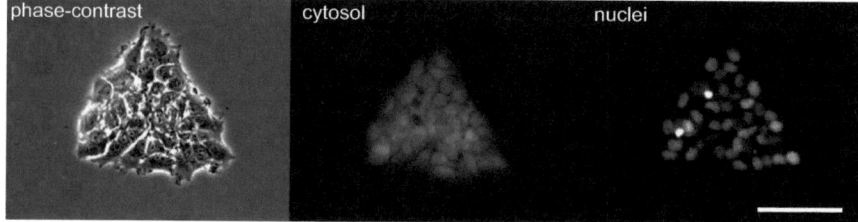

Figure 10.15: Fluorescently labeled MCF-7 cells on a photo-patterned surface. Phase-contrast image (left) of MCF-7 cells expanding from a equilateral triangular pattern with a = 180 μm. Cytosol was labeled with CellTracker Green CMFDA (center) and nuclei were labeled with Hoechst 33342 (right) for live-cell imaging. Image shows still frame from Video A.4.5. Scale bar, 100 μm.

Chapter 11

Discussion

11.1 Collective cell migration on photo-switchable substrates

11.1.1 Epithelial cell sheet expansion

In contrast to traditional scratch assays, the photo-conversion of non-adhesive areas into cell adhesion mediating ones does not release any necrotic factors from disrupted cells and is, therefore, privileged for experiments where such influences are not desired.

Photo-convertible substrates were used to study the impact of initial cluster geometry and incubation time on the expansion of cell clusters that are released from their initial confinement. The results show that there are two size-dependent regimes in MDCK cell sheet expansion: For circular clusters below a certain size of about 5 to 6 cell diameters (r44, r75), all cells participate in the cell sheet expansion in a comparable fashion. Accordingly, the average cell area changes drastically. Once the initial circular clusters exceed that critical size (r104 and larger), cells in the interior of the epithelial sheet play a secondary role for the expansion dynamics which is mainly driven by cells along the cluster boundary. One might speculate that single cell area expansion, initiated by the cells along the cluster boundary, decays in direction to the interior of the cell sheet with a characteristic length of a few cell diameters. Similar lengths scales have been observed in local collective migration behavior within big epithelial cells layers [81, 113].

Not only the general cluster expansion but also the initiation of leader cells depends critically on the size of the initial confinement. For clusters exceeding a certain size (r75) the frequency of observed leader cells f_{LC} is reduced with increasing radius while leader cell formation seems not to be affected if the radius is decreased to r44. Interestingly, the general observation does not change when donut-shaped cell sheets are investigated instead of fully covered circular areas. Thus, one may conclude that once the cell clusters exceed a certain size, it is rather the curvature of the cluster boundary than the total amount of the cells within the cluster that dominates the frequency in leader cell appearance. For smaller cell sheets the different curvature does not affect the leader cell formation, which might be due to the fact that most cells in the cluster are in close proximity to the boundary. The finding of a threshold-like change in leader cell formation coincides with the observation that the change in mean cell area decreases if the size of the initial cluster area is increased from r75 to r104.

By patterning donut-shaped cell clusters and releasing their confinement, one can, for the first time, directly compare the migratory behavior of cells along the concave and the convex boundary of a single sheet. A break in symmetry was detected between these two boundary conditions: Cells at the convex boundary separate into leader and follower cells, whereas cells at the concave boundaries don't. A similar break in symmetry of cell behavior at the outer and inner boundary of a donut-shaped cell sheet has been observed previously. Nelson *et al.* found that proliferation of endothelial and epithelial cells was higher along the outer convex boundary of donut-shaped cell clusters than along the inner concave boundary [48]. This asymmetry in cell behavior is attributed to differences in cell-generated tension within the donut-shaped cell cluster. Cells along the outer boundary apply high tensional forces to the substrate, directed towards the center of the cluster, whereas cells along the inner boundary generate only weak outward-directed forces, suggesting that this asymmetry cause the observed differences in proliferation [48].

Interestingly, the observed geometry-controlled behavior of cell collectives is found in analogy on a single-cell level. Several reports show that the patterning of cells in specific shapes can control various cell functions such as cell polarity, orientation of the cell division plane or cell growth and differentiation [131–134]. Moreover, shape and local curvature of single cells growing on adhesive patterns can determine the spatial appearance of lamellipodia [135] and predefine the direction of cell migration [136]. The fact that lamellipodia protrusions were prominently found at convex cell boundaries but were largely absent from the concave ones of single cells [135] points to an underlying cellular mechanism of the asymmetric cell sheet expansion observed in the present work.

Not only the geometry of the cell sheet, but also the incubation time of the cells affects the expansion characteristic of the cell cluster. The results suggest that collective cell behavior is averted if the cells are incubated for rather short time periods of 9 hours only, instead of 20 hours or more. For such short incubation times with deficient cell-cell interaction, the epithelial cell clusters exhibit an expansion behavior that is typically correlated to single-cell behavior without separation into leader and follower cells. A similar transition from collective to a more individual cell behavior has also been described for endothelial cell clusters in confined geometries after disturbing the cadherins junctions in these cells [48].

In summary, an accurate method is presented to functionalize surfaces with cell adhesion-mediating and switchable inhibiting areas of nearly arbitrary shape. The method is simple and can be used for rapid patterning of large areas and is suitable for high content screening applications. Photo-induced conversion of passivated into cell adhesion-mediating surfaces was used without any mechanical interference with cells at the cluster boundaries. With this method and the experiments it is shown for MDCK cells that expansion of cell sheets of different initial size is mainly controlled by the behavior of cells located along the cluster boundaries. For clusters exceeding a certain size threshold, cells in the interior of the epithelial sheet play a secondary role for the expansion dynamics. More precisely, it is the curvature of the cluster boundary that determines the expansion of these cell sheets. The results imply that mechanical boundary conditions, given by the geometric confinement of the epithelial sheet, can determine leader cell formation and expansion behavior.

Part IV

Conclusions and Outlook

Chapter
12

Cell migration through microfabricated channel structures

12.1 Conclusions

As confirmed by the results presented in sections 6.1–6.3, an *in vitro* model system, the migration chip, has been developed allowing for studying cell migration dynamics in precisely confined 3D channel structures.

Single-cell migration studies with cancer cell lines derived from human pancreas and brain tumors, Panc-1 and MB cells, respectively, showed the versatility of the setup. The migration chip enabled not only to investigate the degree of invasiveness of cancer cells, which correlates with the metastatic potential of the cells, but also to study cell migration dynamics in distinctly confined 3D environments. The setup allows for screening of different channel dimensions at the same time, the surface can be coated with different ECM-derived proteins, and the setup is well-suited for fluorescence live-cell imaging applications.

The development of such microfabricated environments is part of a very active research field with a high potential for industrial applications not only limited to migration studies of cancer cells. This is confirmed by recent publications from other groups, that reported about similar experimental setups and experiments carried out with either cancer cells [104, 137], dendritic cells [103, 138], or neurons [139].

The usefulness of the developed migration chip comprising microfabricated channel structures was demonstrated by investigating the invasive behavior of pancreas and MB cancer cells. In particular, it was found for Panc-1 cells that reorganization of the cytokeratin network, induced by the the phospholipid SPC, facilitated active cell invasion into the channels, but did not affect the migration speeds inside the channels.

12.2 Outlook

There are still many interesting possibilities to further develop and enhance the functionality of channel-based migration assays apart from simply changing the dimensions and shape of the channels:

As already presented and discussed in sections 6.4 and 7.2, patterning of the micro-sized channels with gold nanoparticles will provide more specificity for the bio-functionalization of the channel walls.

In addition to increasing the specificity in surface functionalization, the micro-sized channel structures might be an ideally suited tool to study forces in cell migration. Until now, it is very difficult to measure forces of cells migrating in a 3D environment. If the confined channel geometries would be fabricated with a compliant polymer of stiffnesses in the range of $\sim$0.1-10 kPa, which corresponds to the stiffnesses of native tissue, it would be a perfect model system to study the migration forces via deformation of the channel walls.

Another direction to further improve the versatility of the migration chips would be the incorporation and miniaturization of the migration chips into an automated benchtop device that would allow to directly monitor the migration through the channels with simple optics and a CCD camera, ideally suited for high throughput screenings.

Chapter
13

Cell migration on photo-switchable substrates

13.1 Conclusions

The results in sections 10.1 and 10.2 show that the experimental setup with the photo-switchable surfaced is very well suited for migration studies of cohesive cell sheets. The initial passivation of the surface and and their photo-triggered conversion into a cell adhesive surface is working in a reliable and reproducible way, ideally suited for high content screenings. Large areas of a several mm^2 can be patterned in only a few minutes with arbitrary shapes that are, in principle, only limited by the diffraction of light.

Only a short exposure to UV light without any mechanical or electrical interference is needed to convert the surface properties from being cell repellant to cell adhesive. Thus, the system is eminently suitable for cell studies as the applied dose of UV is not harmful to the cells.

The value of the *in vitro* migration assay was confirmed, as it allowed to investigate expansion dynamics of initially precisely confined cell ensembles. A change in the phenotype of the cell sheet expansion was identified when increasing the radius of circular cell clusters from $r = 75$ μm to $r = 105$ μm.

13.2 Outlook

The next step to further improve the functionalities of the presented model system comprising photo-switchable surfaces would be to move to smaller pattern sizes that are modulated with a confocal laser-scanning setup to guide cell migration *in situ*.

To this point, the surface that remains after the photo-triggered cleavage or the passivating PEG layer does not expose any specific ECM-derived ligands to the cells. It would be interesting and of high experimental value to further enhance the functionalities of this remaining surface. Such enhanced surface functionalizations could be facilitated by using caged compounds whose activities are suppressed by the covalent linkage of photo-cleavable protecting groups, but are restored upon photo-irradiation [140, 141].

Finally, the integration of the photo-active surface functionalization into 3D scaffolds would open the way to fabricate highly functional 3D matrices. The chemical contrast between gold and the PEG hydrogel could be exploited, for example, to selectively bind the photo-cleavable linker to the gold particles to dynamically expose specific ligands on the surface of a 3D matrix.

List of Figures

1.1	Illustration of the wound healing assay	6
1.2	Illustration of the Boyden chamber invasion assay	7
1.3	Illustration of different surface functionalization techniques	8
1.4	Schematic of cell migration in different dimensions	12
2.1	Single steps of cell migration	15
2.2	Cell track from a random walk movement	16
2.3	Cell track with measured time points	17
5.1	Overview of the two-step photolithography and replica molding	29
5.2	Illustaration of tailor-made chrome masks	31
5.3	Experimental approach to fabricate nano-patterned microstructures	36
6.1	Micro-fabricated channel structures	45
6.2	Two different designs of master substrates for migration chips	46
6.3	Assembled migration chips	47
6.4	Interaction of cells with channel structures	48
6.5	SPC-effect on cytoskeleton structure	49
6.6	Summary of Panc-1 cell interactions with the channels	50
6.7	Migration speeds in different environments	52
6.8	Cell migration dynamics inside channels and on lines	53
6.9	Invasive behaviors of MB cells	56

6.10 Photoresist on silicon wafer before and after RIE 59
6.11 Microstructured silicon wafers 60
6.12 Silicon microstructures patterned with gold nanoparticles 61
6.13 Microstructured silicon wafer and hydrogel casting 61

7.1 Schematic of a simple two-component model describing cell invasion 64

9.1 Schematic of the surface functionalization reactions 75
9.2 Illustartion of the working prinicple of photo-switchable surface passivation 77
9.3 Stepwise UV irradiation and cell seeding procedure for experiments with photo-switchable substrates 77
9.4 Illustration of the image-projection exposure setup implemented into an inverted microscope 78

10.1 Phase-contrast images of MDCK cells growing in confined photo-patterned areas 86
10.2 Example of a complex microstructured cell pattern 87
10.3 Example of cell sheet expansion on photo-switchable surface pattern 87
10.4 Cell sheet expansion from a cluster with linear boundaries 88
10.5 Initial cell numbers and cell densities of evaluated cell sheets 89
10.6 Expansion dynamics of circular cell clusters 90
10.7 Quantification of cell sheet expansion dynamics 91
10.8 Frequency of leader cell appearance at circular and donut-shaped patterns of different sizes 93
10.9 Expansion dynamics of donut-shaped cell clusters 95
10.10 Cell behavior is not symmetric for cells along convex and concave boundaries 96
10.11 Leader cell formation is altered by incubation time and curvature . 97
10.12 Immunofluorescence images of MDCK cells seeded on photo-patterned surfaces 99
10.13 Examples of MCF-7 cells seeded on photo-patternened substrates . 100

LIST OF FIGURES 115

10.14 MCF-7 cells expanding into free surface starting from a linear boundary 101

10.15 Fluorescently labeled MCF-7 cells on a photo-patterned surface . . 102

C.1 Synthetic scheme of the photocleavable linker 147

List of Tables

5.1	Photolithography parameters for SU-8 resist	32
5.2	Protocol for microcontact printing	35
5.3	Photolithography parameters for AZ 5214E resist	37
5.4	Anisotropic reactive ion etching parameters	38
5.5	Staining protocol for Panc-1 cells	42
9.1	Staining protocol for MDCK cells	81
B.1	Geometric parameters of photopatterned cell clusters	145
B.2	Summary of leader cell quantification for circular and donut-shaped cell sheets	145
B.3	Summary of leader cell quantification at the inside and outside of donut-shaped cell sheets	146

Bibliography

[1] Jacobsen PF, Jenkyn DJ, Papadimitriou JM (1985) Establishment of a human medulloblastoma cell line and its heterotransplantation into nude mice. *Journal of neuropathology and experimental neurology* 44:472–485.

[2] Soule HD, Vazguez J, Long A, Albert S, Brennan M (1973) A human cell line from a pleural effusion derived from a breast carcinoma. *Journal of the National Cancer Institute* 51:1409–1416.

[3] Madin SH, Darby NB (1958) Established kidney cell lines of normal adult bovine and ovine origin. *Proceedings of the Society for Experimental Biology and Medicine. Society for Experimental Biology and Medicine (New York, N.Y.)* 98:574–576 doi:10.3181/00379727-98-24111.

[4] Lieber M, Mazzetta J, Nelson-Rees W, Kaplan M, Todaro G (1975) Establishment of a continuous tumor-cell line (PANC-1) from a human carcinoma of the exocrine pancreas. *International Journal of Cancer* 15:741–747 doi:10.1002/ijc.2910150505.

[5] Keller R (2002) Shaping the vertebrate body plan by polarized embryonic cell movements. *Science* 298:1950–1954 doi:10.1126/science.1079478.

[6] Sánchez-Madrid F, Pozo M (1999) Leucocyte polarization in cell migration and immune interactions. *The EMBO Journal* 18:501–511 doi:10.1093/emboj/18.3.501.

[7] Martin P (1997) Wound healing – aiming for perfect skin regeneration. *Science* 276:75–81 doi:10.1126/science.276.5309.75.

[8] Yamaguchi H, Wyckoff J (2005) Cell migration in tumors. *Current Opinion in Cell Biology* doi:10.1016/j.ceb.2005.08.002.

[9] Box GM, Eccles SA (2011) Simple experimental and spontaneous metastasis assays in mice. *Methods in molecular biology (Clifton, NJ)* 769:311–329 doi:10.1007/978-1-61779-207-6_21.

[10] He L, Wang X, Montell DJ (2011) Shining light on Drosophila oogenesis: live imaging of egg development. *Current opinion in genetics & development* doi:10.1016/j.gde.2011.08.011.

[11] Yamauchi K, et al. (2006) Development of real-time subcellular dynamic multicolor imaging of cancer-cell trafficking in live mice with a variable-magnification whole-mouse imaging system. *Cancer Research* 66:4208–4214 doi:10.1158/0008-5472.CAN-05-3927.

[12] Condeelis J, Segall JE (2003) Intravital imaging of cell movement in tumours. *Nature reviews Cancer* 3:921–930 doi:10.1038/nrc1231.

[13] Beerling E, Ritsma L, Vrisekoop N, Derksen PWB, van Rheenen J (2011) Intravital microscopy: new insights into metastasis of tumors. *Journal of cell science* 124:299–310 doi:10.1242/jcs.072728.

[14] Le Dévédec SE, et al. (2011) Two-photon intravital multicolour imaging to study metastatic behaviour of cancer cells in vivo. *Methods in molecular biology (Clifton, NJ)* 769:331–349 doi:10.1007/978-1-61779-207-6_22.

[15] Stoletov K, et al. (2010) Visualizing extravasation dynamics of metastatic tumor cells. *Journal of cell science* 123:2332–2341 doi:10.1242/jcs.069443.

[16] Mathias JR, Walters KB, Huttenlocher A (2009) Neutrophil motility in vivo using zebrafish. *Methods in molecular biology (Clifton, NJ)* 571:151–166 doi:10.1007/978-1-60761-198-1_10.

[17] Entschladen F, et al. (2005) Analysis methods of human cell migration. *Experimental Cell Research* 307:418–426 doi:10.1016/j.yexcr.2005.03.029.

[18] Todaro GJ, Lazar GK, Green H (1965) The initiation of cell division in a contact-inhibited mammalian cell line. *Journal of cellular physiology* 66:325–333 doi:10.1002/jcp.1030660310.

[19] Rodriguez LG, Wu X, Guan JL (2005) Wound-healing assay. *Methods in molecular biology (Clifton, NJ)* 294:23–29 doi:10.1385/1-59259-860-9:023.

[20] Yarrow JC, Perlman ZE, Westwood NJ, Mitchison TJ (2004) A high-throughput cell migration assay using scratch wound healing, a comparison of image-based readout methods. *BMC biotechnology* 4:21 doi:10.1186/1472-6750-4-21.

[21] Kam Y, Guess C, Estrada L, Weidow B, Quaranta V (2008) A novel circular invasion assay mimics in vivo invasive behavior of cancer cell lines and distinguishes single-cell motility in vitro. *BMC cancer* 8:198 doi:10.1186/1471-2407-8-198.

[22] Lim JI, Sabouri-Ghomi M, Machacek M, Waterman CM, Danuser G (2010) Protrusion and actin assembly are coupled to the organization of lamellar contractile structures. *Experimental Cell Research* 316:2027–2041 doi:10.1016/j.yexcr.2010.04.011.

[23] Boyden S (1962) The chemotactic effect of mixtures of antibody and antigen on polymorphonuclear leucocytes. *Journal of Experimental Medicine* 115:453–466 doi:10.1084/jem.115.3.453.

[24] Chen HC (2005) Boyden chamber assay. *Methods in molecular biology (Clifton, NJ)* 294:15–22 doi:10.1385/1-59259-860-9:015.

[25] Marshall J (2011) Transwell® invasion assays. *Methods in molecular biology (Clifton, NJ)* 769:97–110 doi:10.1007/978-1-61779-207-6_8.

[26] Geiger B, Bershadsky A, Pankov R, Yamada KM (2001) Transmembrane crosstalk between the extracellular matrix–cytoskeleton crosstalk. *Nature reviews Molecular cell biology* 2:793–805 doi:10.1038/35099066.

[27] Silver FH, Siperko LM, Seehra GP (2003) Mechanobiology of force transduction in dermal tissue. *Skin Research and Technology* 9:3–23 doi:10.1034/j.1600-0846.2003.00358.x.

[28] Kantlehner M, et al. (2000) Surface Coating with Cyclic RGD Peptides Stimulates Osteoblast Adhesion and Proliferation as well as Bone Formation. *Chembiochem : a European journal of chemical biology* 1:107–114 doi:10.1002/1439-7633(20000818)1:2<107::AID-CBIC107>3.0.CO;2-4.

[29] Dicke C, Hähner G (2002) pH-Dependent force spectroscopy of tri(ethylene glycol)- and methyl-terminated self-assembled monolayers adsorbed on gold. *J. Am. Chem. Soc.* 124:12619–12625 doi:10.1021/ja027447n.

[30] Wang RLC, Kreuzer HJ, Grunze M (2011) Molecular Conformation and Solvation of Oligo(ethylene glycol)-Terminated Self-Assembled Monolayers and Their Resistance to Protein Adsorption. *J. Phys. Chem. B* 101:9767–9773 doi:doi: 10.1021/jp9716952 doi: 10.1021/jp9716952.

[31] Israelachvili J, Wennerström H (1996) Role of hydration and water structure in biological and colloidal interactions. *Nature* 379:219–225 doi:10.1038/379219a0.

[32] Doyle AD, Wang FW, Matsumoto K, Yamada KM (2009) One-dimensional topography underlies three-dimensional fibrillar cell migration. *Journal of Cell Biology* 184:481–490 doi:10.1083/jcb.200810041.

[33] Liu VA, Jastromb WE, Bhatia SN (2002) Engineering protein and cell adhesivity using PEO-terminated triblock polymers. *Journal of Biomedical Materials Research* 60:126–134 doi:10.1002/jbm.10005.

[34] Boxshall K, et al. (2006) Simple surface treatments to modify protein adsorption and cell attachment properties within a poly(dimethylsiloxane) microbioreactor. *Surface and Interface Analysis* 38:198–201 doi:10.1002/sia.2274.

[35] Fabrizius-Homan DJ, Cooper SL (1991) A comparison of the adsorption of three adhesive proteins to biomaterial surfaces. *Journal of biomaterials science Polymer edition* 3:27–47.

[36] Hermanson GT (2008) *Bioconjugate Techniques* (Academic Press, Elsevier), 2 edition.

[37] Whitesides GM, Laibinis PE (1990) Wet chemical approaches to the characterization of organic surfaces: self-assembled monolayers, wetting, and the physical-organic chemistry of the solid-liquid interface. *Langmuir* 6:87–96 doi:doi: 10.1021/la00091a013 doi: 10.1021/la00091a013.

[38] Prime K, Whitesides G (1991) Self-assembled organic monolayers: model systems for studying adsorption of proteins at surfaces. *Science* 252:1164–1167 doi:10.1126/science.252.5009.1164.

[39] Love JC, Estroff LA, Kriebel JK, Nuzzo RG, Whitesides GM (2005) Self-Assembled Monolayers of Thiolates on Metals as a Form of Nanotechnology. *Chem. Rev.* 105:1103–1170 doi:doi: 10.1021/cr0300789 doi: 10.1021/cr0300789.

[40] Blodgett KB (1935) Films Built by Depositing Successive Monomolecular Layers on a Solid Surface. *J. Am. Chem. Soc.* 57:1007–1022 doi:doi: 10.1021/ja01309a011 doi: 10.1021/ja01309a011.

[41] Thid D, et al. (2007) Issues of ligand accessibility and mobility in initial cell attachment. *Langmuir* 23:11693–11704 doi:10.1021/la701159u.

[42] Sandrin L, et al. (2010) Cell adhesion through clustered ligand on fluid supported lipid bilayers. *Organic & biomolecular chemistry* 8:1531–1534 doi:10.1039/b924523e.

[43] Andreasson-Ochsner M, et al. (2011) Single cell 3-D platform to study ligand mobility in cell-cell contact. *Lab on a Chip* 11:2876–2883 doi:10.1039/c1lc20067d.

[44] Mrksich M, et al. (1996) Controlling cell attachment on contoured surfaces with self-assembled monolayers of alkanethiolates on gold. *Proc. Nat. Acad. Sci. USA* 93:10775–10778.

[45] Chen CS, Mrksich M, Huang S, Whitesides GM, Ingber DE (1997) Geometric control of cell life and death. *Science* 276:1425–1428 doi:10.1126/science.276.5317.1425.

[46] Younan X, Whitesides GM (1998) Soft lithography. *Annual Review of Materials Science* 28:153–184 doi:10.1146/annurev.matsci.28.1.153.

[47] Brock A, et al. (2003) Geometric determinants of directional cell motility revealed using microcontact printing. *Langmuir* 19:1611–1617 doi:10.1021/la026394k.

[48] Nelson CM, et al. (2005) Emergent patterns of growth controlled by multicellular form and mechanics. *Proc. Nat. Acad. Sci. USA* 102:11594–11599 doi:10.1073/pnas.0502575102.

[49] Théry M, Piel M (2009) Adhesive micropatterns for cells: a microcontact printing protocol. *Cold Spring Harbor protocols* 2009:pdb.prot5255 doi:10.1101/pdb.prot5255.

[50] Azioune A, Storch M, Bornens M, Théry M, Piel M (2009) Simple and rapid process for single cell micro-patterning. *Lab on a Chip* 9:1640–1642 doi:10.1039/b821581m.

[51] Möller M, Spatz JP, Roescher A (1996) Gold nanoparticles in micellar poly(styrene)-b-poly(ethylene oxide) films—size and interparticle distance control in monoparticulate films. *Advanced Materials* 8:337–340 doi:10.1002/adma.19960080411.

[52] Spatz JP, Mößmer S, Möller M (1996) Mineralization of Gold Nanoparticles in a Block Copolymer Microemulsion. *Chemistry - A European Journal* 2:1552–1555 doi:10.1002/chem.19960021213.

[53] Lohmüller T, et al. (2011) Nanopatterning by block copolymer micelle nanolithography and bioinspired applications. *Biointerphases* 6:MR1–12 doi:10.1116/1.3536839.

[54] Aragüés B (2011) Ph.D. thesis (Ruprecht-Karls Universität Heidelberg).

[55] Teixeira AI, Abrams GA, Bertics PJ, Murphy CJ, Nealey PF (2003) Epithelial contact guidance on well-defined micro- and nanostructured substrates. *Journal of cell science* 116:1881–1892 doi:10.1242/jcs.00383.

[56] Motlagh D, Senyo SE, Desai TA, Russell B (2003) Microtextured substrata alter gene expression, protein localization and the shape of cardiac myocytes. *Biomaterials* 24:2463–2476 doi:10.1016/S0142-9612(02)00644-0.

[57] Miyoshi H, et al. (2010) Control of highly migratory cells by microstructured surface based on transient change in cell behavior. *Biomaterials* 31:8539–8545 doi:10.1016/j.biomaterials.2010.07.076.

[58] Ghibaudo M, et al. (2009) Substrate Topography Induces a Crossover from 2D to 3D Behavior in Fibroblast Migration. *Biophysical journal* 97:357–368 doi:10.1016/j.bpj.2009.04.024.

[59] Wolf K, et al. (2007) Multi-step pericellular proteolysis controls the transition from individual to collective cancer cell invasion. *Nat Cell Biol* 9:893–904 doi:10.1038/ncb1616.

[60] Wolf K, et al. (2009) Collagen-based cell migration models in vitro and in vivo. *Seminars in cell & developmental biology* 20:931–941 doi:10.1016/j.semcdb.2009.08.005.

[61] Lämmermann T, et al. (2008) Rapid leukocyte migration by integrin-independent flowing and squeezing. *Nature* 453:51–55 doi:10.1038/nature06887.

[62] Lutolf MP, Raeber GP, Zisch AH, Tirelli N, Hubbell JA (2003) Cell-Responsive Synthetic Hydrogels. *Advanced Materials* 15:888–892 doi:10.1002/adma.200304621.

[63] Liu Tsang V, Bhatia SN (2004) Three-dimensional tissue fabrication. *Advanced Drug Delivery Reviews* 56:1635–1647 doi:10.1016/j.addr.2004.05.001.

[64] Herrmann H, Hesse M, Reichenzeller M, Aebi U, Magin TM (2003) Functional complexity of intermediate filament cytoskeletons: from structure to assembly to gene ablation. *International review of cytology* 223:83–175 doi:10.1016/S0074-7696(05)23003-6.

[65] Herrmann H, Strelkov SV, Burkhard P, Aebi U (2009) Intermediate filaments: primary determinants of cell architecture and plasticity. *The Journal of clinical investigation* 119:1772–1783 doi:10.1172/JCI38214.

[66] Rowat AC, Lammerding J, Herrmann H, Aebi U (2008) Towards an integrated understanding of the structure and mechanics of the cell nucleus. *Bioessays* 30:226–236 doi:10.1002/bies.20720.

[67] Desai A, Mitchison TJ (1997) Microtubule polymerization dynamics. *Annual Review of Cell and Developmental Biology* 13:83–117 doi:10.1146/annurev.cellbio.13.1.83.

[68] Howard J, Hyman AA (2003) Dynamics and mechanics of the microtubule plus end. *Nature* 422:753–758 doi:10.1038/nature01600.

[69] Friedl P (2004) Prespecification and plasticity: shifting mechanisms of cell migration. *Current Opinion in Cell Biology* 16:14–23 doi:10.1016/j.ceb.2003.11.001.

[70] Lauffenburger DA, Horwitz AF (1996) Cell migration: a physically integrated molecular process. *Cell* 84:359–369 doi:10.1016/S0092-8674(00)81280-5.

[71] Friedl P, Wolf K (2003) Tumour-cell invasion and migration: diversity and escape mechanisms. *Nature reviews Cancer* 3:362–374 doi:10.1038/nrc1075.

[72] Palecek SP, Loftus JC, Ginsberg MH, Lauffenburger DA, Horwitz AF (1997) Integrin-ligand binding properties govern cell migration speed through cell-substratum adhesiveness. *Nature* 385:537–540 doi:10.1038/385537a0.

[73] Dickinson RB, Tranquillo RT (1993) Optimal estimation of cell movement indices from the statistical analysis of cell tracking data. *AIChE Journal* 39:1995–2010 doi:10.1002/aic.690391210.

[74] Codling EA, Plank MJ, Benhamou S (2008) Random walk models in biology. *Journal of the Royal Society, Interface* 5:813–834 doi:10.1098/rsif.2008.0014.

[75] Wirtz D, Konstantopoulos K, Searson PC (2011) The physics of cancer: the role of physical interactions and mechanical forces in metastasis. *Nature reviews Cancer* 11:512–522 doi:10.1038/nrc3080.

[76] Engler AJ, Sen S, Sweeney HL, Discher DE (2006) Matrix elasticity directs stem cell lineage specification. *Cell* 126:677–689 doi:10.1016/j.cell.2006.06.044.

[77] Paszek MJ, et al. (2005) Tensional homeostasis and the malignant phenotype. *Cancer cell* 8:241–254 doi:10.1016/j.ccr.2005.08.010.

[78] Wang JHC, Lin JS (2007) Cell traction force and measurement methods. *Biomechanics and modeling in mechanobiology* 6:361–371 doi:10.1007/s10237-006-0068-4.

[79] Delanoë-Ayari H, Rieu JP, Sano M (2010) 4D traction force microscopy reveals asymmetric cortical forces in migrating Dictyostelium cells. *Physical Review Letters* 105:248103 doi:10.1103/PhysRevLett.105.248103.

[80] Maskarinec SA, Franck C, Tirrell DA, Ravichandran G (2009) Quantifying cellular traction forces in three dimensions. *Proc. Nat. Acad. Sci. USA* 106:22108–22113 doi:10.1073/pnas.0904565106.

[81] Tambe DT, et al. (2011) Collective cell guidance by cooperative intercellular forces. *Nature Materials* 10:469–475 doi:10.1038/nmat3025.

[82] Munevar S, Wang YL, Dembo M (2001) Distinct roles of frontal and rear cell-substrate adhesions in fibroblast migration. *Molecular biology of the cell* 12:3947–3954.

[83] Ananthakrishnan R, Ehrlicher A (2007) The forces behind cell movement. *International journal of biological sciences* 3:303–317.

[84] Kraning-Rush CM, Carey SP, Califano JP, Smith BN, Reinhart-King CA (2011) The role of the cytoskeleton in cellular force generation in 2D and 3D environments. *Physical biology* 8:015009 doi:10.1088/1478-3975/8/1/015009.

[85] Legant WR, et al. (2010) Measurement of mechanical tractions exerted by cells in three-dimensional matrices. *Nature Methods* 7:969–971 doi:10.1038/nmeth.1531.

[86] Hawkins RJ, et al. (2009) Pushing off the Walls: A Mechanism of Cell Motility in Confinement. *Physical Review Letters* 102:4 doi:10.1103/PhysRevLett.102.058103.

[87] Sahai E (2007) Illuminating the metastatic process. *Nature reviews Cancer* 7:737–749 doi:10.1038/nrc2229.

[88] Kumar S, Weaver VM (2009) Mechanics, malignancy, and metastasis: the force journey of a tumor cell. *Cancer metastasis reviews* 28:113–127 doi:10.1007/s10555-008-9173-4.

[89] Weinberg RA (2007) The Biology of Cancer. *Garland Science*.

[90] Seufferlein T, Rozengurt E (1995) Sphingosylphosphorylcholine rapidly induces tyrosine phosphorylation of p125FAK and paxillin, rearrangement of the actin cytoskeleton and focal contact assembly. Requirement of p21rho in the signaling pathway. *The Journal of biological chemistry* 270:24343–24351 doi:10.1074/jbc.270.41.24343.

[91] Beil M, et al. (2003) Sphingosylphosphorylcholine regulates keratin network architecture and visco-elastic properties of human cancer cells. *Nat Cell Biol* 5:803–811 doi:10.1038/ncb1037.

[92] Suresh S, et al. (2005) Connections between single-cell biomechanics and human disease states: gastrointestinal cancer and malaria. *Acta Biomaterialia* 1:15–30 doi:10.1074/jbc.270.41.24343.

[93] Tserepi A, Gogolides E, Cardinaud C, Rolland L, Turban G (1998) Highly anisotropic silicon and polysilicon room-temperature etching using fluorine-based high density plasmas. *Microelectronic Engineering* 41-42:411–414 doi:10.1016/S0167-9317(98)00095-1.

[94] Milenin A, Jamois C, Geppert T, Gösele U, Wehrspohn R (2005) SOI planar photonic crystal fabrication: Etching through SiO2/Si/SiO2 layer systems using fluorocarbon plasmas. *Microelectronic Engineering* 81:15–21 doi:10.1016/j.mee.2005.02.007.

[95] Busch T (2008) Ph.D. thesis (Medizinische Fakultät der Universität Ulm).

[96] Rasband W (1997) ImageJ. *U. S. National Institutes of Health, Bethesda, Maryland, USA* http://rsb.info.nih.gov/ij/.

[97] Polkinghorn WR, Tarbell NJ (2007) Medulloblastoma: tumorigenesis, current clinical paradigm, and efforts to improve risk stratification. *Nature Clinical Practice Oncology* 4:295–304 doi:10.1038/ncponc0794.

[98] Lu J, et al. (2005) MicroRNA expression profiles classify human cancers. *Nature* 435:834–838 doi:10.1038/nature03702.

[99] He L, et al. (2005) A microRNA polycistron as a potential human oncogene. *Nature* 435:828–833 doi:10.1038/nature03552.

[100] Xu S, Witmer PD, Lumayag S, Kovacs B, Valle D (2007) MicroRNA (miRNA) transcriptome of mouse retina and identification of a sensory organ-specific miRNA cluster. *The Journal of biological chemistry* 282:25053–25066 doi:10.1074/jbc.M700501200.

[101] Bai AH, et al. MicroRNA-182 promotes metastatic dissemination of non-sonic hedgehog-medulloblastoma. *Acta Neuropathologica*, submitted.

[102] Irimia D, Charras G, Agrawal N, Mitchison T, Toner M (2007) Polar stimulation and constrained cell migration in microfluidic channels. *Lab on a Chip* 7:1783–1790 doi:10.1039/b710524j.

[103] Faure-André G, et al. (2008) Regulation of dendritic cell migration by CD74, the MHC class II-associated invariant chain. *Science* 322:1705–1710 doi:10.1126/science.1159894.

[104] Irimia D, Toner M (2009) Spontaneous migration of cancer cells under conditions of mechanical confinement. *Integrative Biology* 1:506–512 doi:10.1039/b908595e.

[105] Vereycken V, Gruler H, Bucherer C, Lacombe C, Lelièvre J (1995) The linear motor in the human neutrophil migration. *Journal De Physique Iii* 5:1469–1480 doi:10.1051/jp3:1995204.

[106] Nieminen M, et al. (2006) Vimentin function in lymphocyte adhesion and transcellular migration. *Nat Cell Biol* 8:156–162 doi:10.1038/ncb1355.

[107] Pouthas F, et al. (2008) In migrating cells, the Golgi complex and the position of the centrosome depend on geometrical constraints of the substratum. *Journal of cell science* 121:2406–2414 doi:10.1242/jcs.026849.

[108] van der Meer AD, Vermeul K, Poot AA, Feijen J, Vermes I (2010) A microfluidic wound-healing assay for quantifying endothelial cell migration. *Amer-*

ican journal of physiology. Heart and circulatory physiology 298:H719–25 doi:10.1152/ajpheart.00933.2009.

[109] Keese CR, Wegener J, Walker SR, Giaever I (2004) Electrical wound-healing assay for cells in vitro. *Proc. Nat. Acad. Sci. USA* 101:1554–1559 doi:10.1073/pnas.0307588100.

[110] Fischer EG, Stingl A, Kirkpatrick CJ (1990) Migration assay for endothelial cells in multiwells. Application to studies on the effect of opioids. *Journal of Immunological Methods* 128:235–239 doi:10.1016/0022-1759(90)90215-H.

[111] Folch A, Jo BH, Hurtado O, Beebe DJ, Toner M (2000) Microfabricated elastomeric stencils for micropatterning cell cultures. *Journal of Biomedical Materials Research* 52:346–353 doi:10.1002/1097-4636(200011)52:2<346::AID-JBM14>3.0.CO;2-H.

[112] Ostuni E, Kane R, Chen C, Ingber D, Whitesides G (2000) Patterning mammalian cells using elastomeric membranes. *Langmuir* 16:7811–7819 doi:10.1021/la000382m.

[113] Poujade M, et al. (2007) Collective migration of an epithelial monolayer in response to a model wound. *Proc. Nat. Acad. Sci. USA* 104:15988–15993 doi:10.1073/pnas.0705062104.

[114] Doran MR, Mills RJ, Parker AJ, Landman KA, Cooper-White JJ (2009) A cell migration device that maintains a defined surface with no cellular damage during wound edge generation. *Lab on a Chip* 9:2364–2369 doi:10.1039/b900791a.

[115] Xingyu J, Ferrigno R, Mrksich M, Whitesides GM (2003) Electrochemical desorption of self-assembled monolayers noninvasively releases patterned cells from geometrical confinements. *J. Am. Chem. Soc.* 125:2366–2367 doi:10.1021/ja029485c.

[116] Yousaf MN, Houseman BT, Mrksich M (2001) Using electroactive substrates to pattern the attachment of two different cell populations. *Proc. Nat. Acad. Sci. USA* 98:5992–5996 doi:10.1073/pnas.101112898.

[117] Raghavan S, Desai RA, Kwon Y, Mrksich M, Chen CS (2010) Micropatterned dynamically adhesive substrates for cell migration. *Langmuir* 26:17733–17738 doi:10.1021/la102955m.

[118] Kikuchi Y, et al. (2008) Grafting Poly(ethylene glycol) to a Glass Surface via a Photocleavable Linker for Light-induced Cell Micropatterning and Cell Proliferation Control. *Chemistry Letters* 37:1062–1063 doi:10.1246/cl.2008.1062.

[119] Nakayama H, et al. (2010) Silane coupling agent bearing a photoremovable succinimidyl carbonate for patterning amines on glass and silicon surfaces with controlled surface densities. *Colloids and Surfaces B: Biointerfaces* 76:88–97 doi:10.1016/j.colsurfb.2009.10.020.

[120] Fong E, Tzlil S, Tirrell DA (2010) Boundary crossing in epithelial wound healing. *Proc. Nat. Acad. Sci. USA* 107:19302–19307 doi:10.1073/pnas.1008291107.

[121] Angelini TE, Hannezo E, Trepat X, Fredberg JJ, Weitz DA (2010) Cell migration driven by cooperative substrate deformation patterns. *Physical Review Letters* 104:168104 doi:10.1103/PhysRevLett.104.168104.

[122] Huergo M, Pasquale M, González P, Bolzán A, Arvia A (2011) Dynamics and morphology characteristics of cell colonies with radially spreading growth fronts. *Physical Review E* 84 doi:10.1103/PhysRevE.84.021917.

[123] Omelchenko T, Vasiliev JM, Gelfand IM, Feder HH, Bonder EM (2003) Rho-dependent formation of epithelial "leader" cells during wound healing. *Proc. Nat. Acad. Sci. USA* 100:10788–10793 doi:10.1073/pnas.1834401100.

[124] Khalil AA, Friedl P (2010) Determinants of leader cells in collective cell migration. *Integrative Biology* 2:568–574 doi:10.1039/c0ib00052c.

[125] Nakanishi J, et al. (2004) Photoactivation of a substrate for cell adhesion under standard fluorescence microscopes. *J. Am. Chem. Soc.* 126:16314–16315 doi:10.1021/ja044684c.

[126] Kaneko S, et al. (2011) Photocontrol of cell adhesion on amino-bearing surfaces by reversible conjugation of poly(ethylene glycol) via a photocleavable linker. *Physical Chemistry Chemical Physics* 13:4051 doi:10.1039/c0cp02013c.

[127] Martin P, Lewis J (1992) Actin cables and epidermal movement in embryonic wound healing. *Nature* 360:179–183 doi:10.1038/360179a0.

[128] Danjo Y, Gipson IK (1998) Actin 'purse string' filaments are anchored by E-cadherin-mediated adherens junctions at the leading edge of the epithelial wound, providing coordinated cell movement. *Journal of cell science* 111 (Pt 22):3323–3332.

[129] Martin P, Parkhurst SM (2004) Parallels between tissue repair and embryo morphogenesis. *Development (Cambridge, England)* 131:3021–3034 doi:10.1242/dev.01253.

[130] Florian P, Schöneberg T, Schulzke JD, Fromm M, Gitter AH (2002) Single-cell epithelial defects close rapidly by an actinomyosin purse string mechanism with functional tight junctions. *The Journal of physiology* 545:485–499 doi:10.1113/jphysiol.2002.031161.

[131] Desai RA, Gao L, Raghavan S, Liu WF, Chen CS (2009) Cell polarity triggered by cell-cell adhesion via E-cadherin. *Journal of cell science* 122:905–911 doi:10.1242/jcs.028183.

[132] Dupin I, Camand E, Etienne-Manneville S (2009) Classical cadherins control nucleus and centrosome position and cell polarity. *The Journal of Cell Biology* 185:779–786 doi:10.1083/jcb.200812034.

[133] Théry M, et al. (2005) The extracellular matrix guides the orientation of the cell division axis. *Nature cell biology* 7:947–953 doi:10.1038/ncb1307.

[134] Théry M (2010) Micropatterning as a tool to decipher cell morphogenesis and functions. *Journal of cell science* 123:4201–4213 doi:10.1242/jcs.075150.

[135] James J, Goluch ED, Hu H, Liu C, Mrksich M (2008) Subcellular curvature at the perimeter of micropatterned cells influences lamellipodial distribution and cell polarity. *Cell Motility and the Cytoskeleton* 65:841–852 doi:10.1002/cm.20305.

[136] Xingyu J, Bruzewicz DA, Wong AP, Piel M, Whitesides GM (2005) Directing cell migration with asymmetric micropatterns. *Proc. Nat. Acad. Sci. USA* 102:975–978 doi:10.1073/pnas.0408954102.

[137] Wolfer A, et al. (2010) MYC regulation of a "poor-prognosis" metastatic cancer cell state. *Proc. Nat. Acad. Sci. USA* doi:10.1073/pnas.0914203107.

[138] Heuzé ML, Collin O, Terriac E, Lennon-Duménil AM, Piel M (2011) Cell migration in confinement: a micro-channel-based assay. *Methods in molecular biology (Clifton, NJ)* 769:415–434 doi:10.1007/978-1-61779-207-6_28.

[139] Kim YT, Karthikeyan K, Chirvi S, Davé DP (2009) Neuro-optical microfluidic platform to study injury and regeneration of single axons. *Lab on a Chip* 9:2576–2581 doi:10.1039/b903720a.

[140] Ellis-Davies GCR (2007) Caged compounds: photorelease technology for control of cellular chemistry and physiology. *Nature Publishing Group* 4:619–628 doi:10.1038/nmeth1072.

[141] Nakanishi J, et al. (2009) Light-Regulated Activation of Cellular Signaling by Gold Nanoparticles That Capture and Release Amines. *J. Am. Chem. Soc.* 131:3822–3823 doi:10.1021/ja809236a.

Appendix
A

Overview of attached Videos

The following videos can be found on the attached DVD.

A.1 Panc-1 cells in microchannels and on 1D lines

A.1.1 Panc-1_1_1.avi

Panc-1 cells seeded in close proximity to channels with a width of 15 μm. Cells migrate on the flat surface in front of the channels and are able to permeate the channels without major deformation. Channel dimensions are 15x11x150 μm (WxHxL); duration 16:00 h; scale bar, 50 μm.

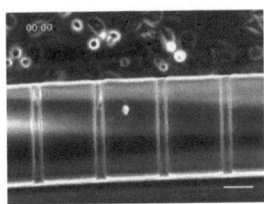

A.1.2 Panc-1_1_2.avi

Representative examples of Panc-1 cells seeded in close proximity to the channels that show penetrative, invasive, and permeative behavior (from left to right).

Channel dimensions are 7x11x150 μm (WxHxL); duration 6:34 h; scale bar, 50μm.

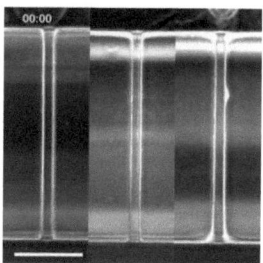

A.1.3 Panc-1_1_3.avi

P1 cells with fluorescently labeled keratin 8 and 18 migrating through a channel. Channel dimensions are 15x11x150 μm (WxHxL); duration 16:00 h; scale bar, 50 μm.

A.1.4 Panc-1_1_4avi

Panc-1 cells migrating along μCP lines of fibronectin on a non-adhesive surface. Line width is 7 μm; duration 6:34 h; scale bar, 50 μm.

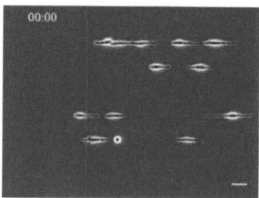

A.2 MB cells in microchannels

A.2.1 MB_2_1.avi

DAOY cells stably transfected with miR-182 were seeded closely to the channels. The third channel on the right hand side shows a representative example of a permeating cell. Duration 23:50 h; scale bar, 100 μm.

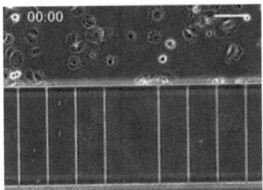

A.2.2 MB_2_2.avi

DAOY cells stably transfected with empty pCMX vector. All four channels on the right hand side showed the typical penetrative behavior where cells continuously try to enter the channels throughout the whole experiment, while the nuclei are not able to enter the channels. Duration 23:50 h; scale bar, 100 μm.

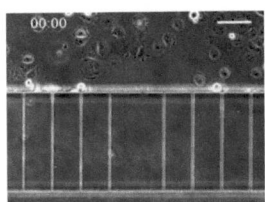

A.2.3 MB_2_3.avi

DAOY cells with miR-182 knockdown after 24h. Less invasive activity was observed and cells were only trying to enter the channels for short periods. Duration 23:50 h; scale bar, 100μm.

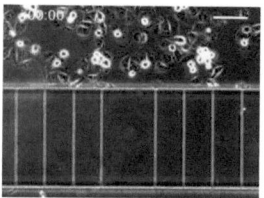

A.2.4 MB_2_4.avi

DAOY cells treated with scrambled siRNA. A DAOY cell invaded into the channel and migrated until the middle of the channel. Then the cell returned to the starting point. This special invasion behavior could be found in the second channel on right hand side. Duration 23:50 h; scale bar, 100 μm.

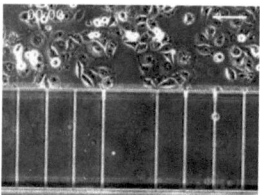

A.2.5 MB_2_5.avi

Med8A cells stably transfected with miR-182. Most of the cells remained in the same position, and did not enter the channel. The fourth channel on the left side showed a rare example for a single invasion. Duration 23:20 h; scale bar, 100 μm.

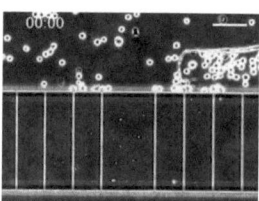

A.2.6 MB_2_6.avi

Med8A cells stably transfected with empty pCMX vector. All cells included in the field of view remained in a similar position and no invasive behaviors were observed at all. Duration 23:50 h; scale bar, 100 μm.

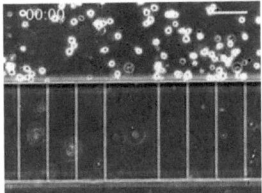

A.3 MDCK cells on photo-active substrates

A.3.1 MDCK_3_1.avi

Time-course of a typical migration experiment. Video shows MDCK cells seeded on a photo-patterned surface and incubated for 38 h. Starting point is 30 min before the cells were released from their confinement by irradiating the whole surface with a short UV exposure. After a short delay, cells start to expand towards the outside as well as they are closing the inner hole of the donut structure. Pattern r142/81; 38 h incubation time; duration 4:45 h; scale bar, 100 μm.

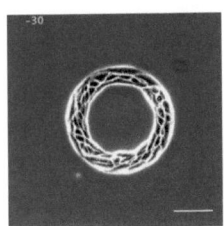

A.3.2 MDCK_3_2.avi

Cell sheet expanding from a stripe-shaped cell cluster with initially almost straight boundries and a width of 250 μm; video starts 20 h after cell seeding; duration

16:00 h; scale bar, 100 μm.

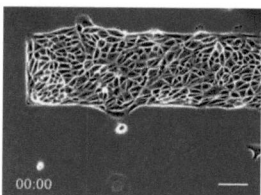

A.3.3 MDCK_3_3.avi

Patterning of complex cell clusters. Cells growing in the pattern of the Max Planck Society logo; video starts 26 h after cell seeding, confinement was released after the first 35 min of the video; duration 12:25 h; scale bar, 500 μm.

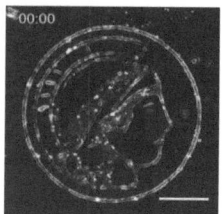

A.3.4 MDCK_3_4.avi

Expansion of circular cell clusters after different incubation times. Cells have been seeded for 9 h and 25 h on circular photopatterned substrates. Videos start directly after releasing the confinement. From the cell cluster incubated for 9 h more leader cells are evolving, cells inside the cell sheet are more motile and the mean cluster radius is expanding faster then for the 25 h case (see Figure 10.7). Pattern r104; flood exposure after 9 h and 25 h incubation time; duration 5:00 h; scale bar, 100 μm.

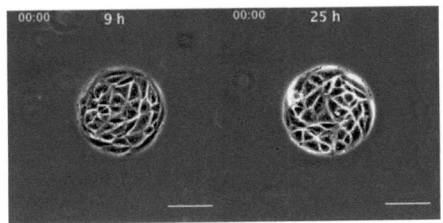

A.3.5 MDCK_3_5.avi

Influence of curvature on boundary cell behavior. Video shows a cell cluster with a donut shaped pattern while being incubated under normal cell culture conditions. Cells adapt to the geometrical confinement and show active lamellar protrusions mostly towards the outside of the pattern. Hardly any protrusion activity is found along the inner boundary. The prominent halo-effect along the inner boundary caused by the phase-contrast illumination indicates that the cell membranes are higher along the inner boundary compared to the thin and flat protrusions along the outside. Pattern r142/81; video starts 9 h after cell seeding; duration 5:00 h; scale bar, 100 μm.

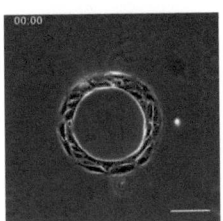

A.3.6 MDCK_3_6.avi

Angular resolved protrusion analysis. Cell cluster outlines have been detected and assigned to polar coordinates whose radial displacement has been tracked over time. The left part shows an overlay of the phase-contrast video and the cluster outlines with their assigned angular positions and the left part shows the color coded accumulation of all outlines and positions over time. Pattern r173/120;

video starts 24 h after cell seeding; duration 2:00 h; scale bar, 100 μm.

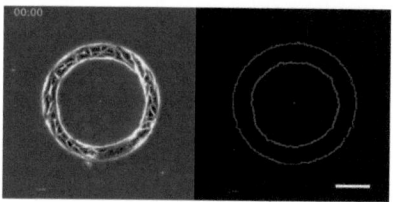

A.3.7 MDCK_3_7.avi

Circular cell clusters with cells moving in a coordinated fashion inside the confinement. Pattern r75; video starts 20 h after cell seeding; duration 8:40 h; scale bar, 100 μm.

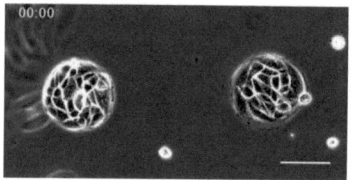

A.4 MCF-7 cells on photo-active subastrates

A.4.1 MCF-7_4_1.avi

Linear boundary of MCF-7 cell sheet expanding into the free area. No pronounced separation of cells into leader and follower cells is observed. Video starts 20 h after cell seeding; duration 12:00 h; scale bar, 100 μm.

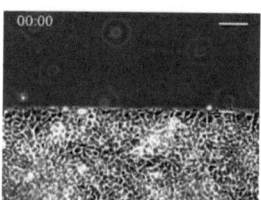

A.4.2 MCF-7_4_2.avi

Cell clusters of MCF-7 cell in the shape of equilateral triangles with a = 180 μm expanding after confinement was released; video starts 24 h after cell seeding; duration 12:30 h; scale bar, 100 μm.

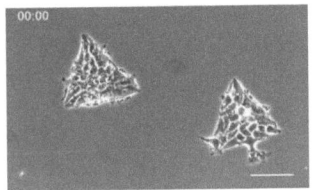

A.4.3 MCF-7_4_3.avi

Donut-shaped cell cluster expanding to the outside and inside after confinement was released. The inner part of the cluster is completely closed after 7 h of cell expansion. Pattern r140/60; video starts 20 h after cell seeding; duration 12:00 h; scale bar, 100 μm.

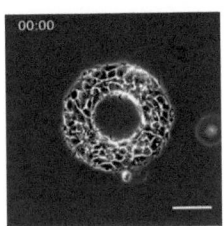

A.4.4 MCF-7_4_4.avi

Cell sheet expansion of stripe-shaped cell cluster after photo-release of the confinement. Unlike MDCK cells, the border of MCF-7 cell sheet is expanding the most in regions of highest initial cell density, without separation into leader and follower cells. 160 μm thick stripe; video starts 20 h after cell seeding; duration 24:00 h; scale bar, 100 μm.

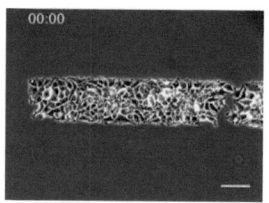

A.4.5 MCF-7_4_5.avi

Phase-contrast (left) and fluorescence live-cell video of MCF-7 with the cytosol (center, CellTracker Green CMFDA) and nuclei (right, Hoechst 33342) labeled. Images were recorded at a frame rate of 5 min/frame (phase-contrast), 15 min/frame (CMFDA) and 30 min/frame (Hoechst) Equilateral triangle with a = 180 μm; video starts 24 h after cell seeding; duration 12:00 h; scale bar, 100 μm.

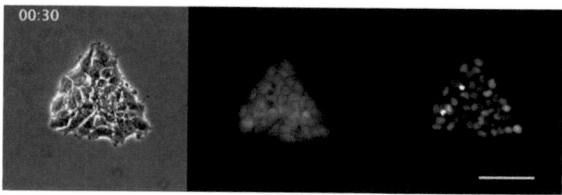

Appendix B

Additional Tables

Table B.1: Geometric parameters of photopatterned cell clusters. Circular and donut-shaped clusters that were used to study collective cell migration.

shape	circle	circle	circle	circle	donut (out; in)	donut (out; in)
radius (μm)	44	75	104	142	142; 81	173; 120
curvature ($*10^{-3}\mu m^{-1}$)	22.9	13.4	9.6	7.1	7.1; 12.7	5.8; 8.3
area ($*10^3 \mu m^2$)	6	17.5	34.4	63.2	43.7	94.0
perimeter (μm)	275	469	657	891	891; 508	1037; 754

Table B.2: Summary of leader cell quantification for circular and donut-shaped cell sheets.

pattern	r44	r75	r104	r142	r142/81	r173/120
median of counted leader cells	3	5	5	6	6	5.5
median frequency of leader cell appearance f_{LC}	1.09	1.07	0.76	0.67	0.67	0.51

Table B.3: Summary of leader cell quantification at the inside and outside of donut-shaped cell sheets. Cell clusters were incubated for different times as indicated prior to confinement release.

pattern	\multicolumn{6}{c}{r142/81}					
incubation time	9 h		25 h		38 h	
donut boundary	in	out	in	out	in	out
median of counted leader cells	7	13	1	6	2	6
median frequency of leader cell appearance f_{LC}	1.38	1.45	0.20	0.67	0.39	0.67

Appendix C

Synthesis of photocleavable linker

The photocleavable linker 1-[5-methoxy-2-nitro-4-(3-trimethoxysilylpropyloxy)-phenyl]-ethyl N-succinimidyl carbonate (**5**) was provided by our collaboration partner Professor Kazuo Yamaguchi at the Kanagawa University, Japan. As depicted in the synthetic scheme C.1, it was synthesized in four steps starting with the commercially available 4-hydroxy-5-methoxy-2-nitroacetophenone (**1**) [118, 119]. All chemicals used, if not indicated differently, were purchased either from Wako (Japan), TCI (Japan) or Sigma–Aldrich (USA).

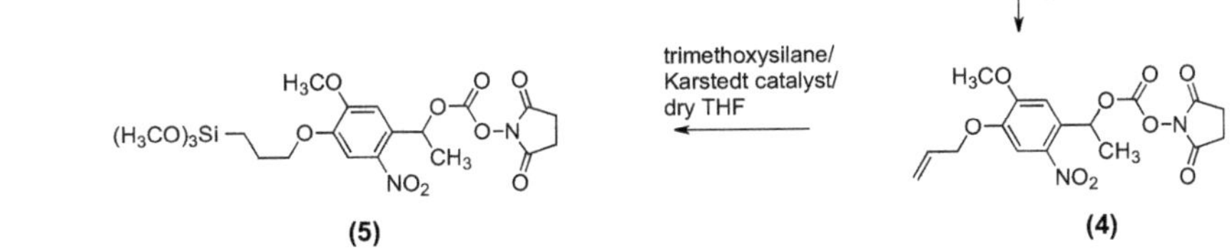

Figure C.1: **Synthetic scheme of the photocleavable linker.**

Allyl bromide (2.4 mL, 28 mmol) is added to a solution of (**1**) (5.0 g, 24 mmol) and potassium carbonate (3.9 g, 28 mmol) in 200 mL dry acetonitrile.

Stirring at 80°C for 3 hours under nitrogen atmosphere, concentrating the solution, adding 300 mL water and 20 mL 2 M HCl, the product 4-Allyloxy-5-methoxy-2-nitroacetophenone (**2**) is being extracted with chlorophorm, dried over MgSO$_4$ and after removing the solvent a yellowish-white solid is obtained with a yield of 99% (5.9 g, 24 mmol).

In the next step the carbonyl group was reduced to a tertiary alcohol by stirring a solution of (**2**) (5.8g, 23 mmol) for 30 min in an ice bath with sodium tetrayhdroborate (2.6 g, 69 mmol), 50 mL tetrahydrofuran (THF) and 100 mL methanol and stirred for another 3 hours at r.t.. Concentrating the solution, adding 100 mL water and 20 mL 2 M HCl, extracting with 3x150 mL chlorophorm, drying over MgSO$_4$ and removing the solvent lead to the 1-(4-Allyloxyl-5-methoxy-2-nitrophenyl)ethanol (**3**) product as a yellowish white solid with a yield of 97% (5.7 g, 22 mmol).

The N-succinimidyl carbamate, which allows for the coupling of primary amines, was obtained by adding N,N-disuccinimidyl carbonate (3.7 g, 14 mmol) to a solution of (**3**) (1.8 g, 7.0 mmol) in acetonitril in the presence of 4 mL triethylamine. Stirring for 6.5 hours at r.t. under nitrogen atmosphere, concentrating the solution, adding 50 mL water and 10 mL 2 M HCl, extracting with chlorophorm, washing the organic layer with saturated NaHCO$_3$, drying over MgSO$_4$ and removing the solvent lead to the 1-(4-Allyloxy-5-methoxy-2-nitrophenyl)ethyl N-succinimidyl carbonate (**4**) product as a yellowish white solid with a yield of 95% (2.6 g, 6.7 mmol).

Finally, the silane group was introduced by adding trimethoxysilane (0.26 g, 2.1 mmol) in the presence of Karstedt's catalyst (10 drops) to a solution of (**4**) (0.26 g, 0.66 mmol) in 8 mL dry THF. Stirring the solution at r.t. for 1 hour under nitrogen atmosphere, concentrating the crude product, and purifying it by silica gel column chromatography (hexane/ethylacetate/tetramethoxysilane, 50:50:1) lead to the final product (**5**) as a yellow, viscous compound with a yield of 50% (0.17 g, 0.33 mmol).

Appendix D

MATLAB codes

D.1 Binarization of phase-contrast images

Starting parameters were the following: fudgeFactor = 0.25 (the smaller the value, the more objects are detected); strelSizeline = 5; strelSizedisk = 10; thinsteps = 3; smoothsteps = 1; strelSizediam = 2.

```matlab
function [BW] = PhSegment02(I,fudgeFactor,strelSizeline, ...
    strelSizedisk, strelSizediam, thinsteps, smoothsteps)
%% Calculate gradient image and threshold; sobel edge detection
[junk threshold] = edge(I, 'sobel');
BWs = edge(I,'sobel', threshold * fudgeFactor);

%% Increase the detected regions by dilating
se90 = strel('line', strelSizeline, 90);
se0 = strel('line', strelSizeline, 0);
BWsdil = imdilate(BWs, [se90 se0]);

%% Fill all the smaller holes, applying rel. big STREL disk
se = strel('disk', strelSizedisk);
BWpartfill = imclose(BWsdil, se);

%% Make the partfill object smaller to preserve thin
    %boundary shapes
BWthin = bwmorph(BWpartfill, 'thin', thinsteps);

%% Overlay partly filled object with BWsdil
BWrough=BWsdil;BWrough(BWthin)=1;

%% Smoothen boundaries
seD    = strel('diamond',strelSizediam);
```

```
24  BWsmooth =BWrough;
25  for i=1:smoothsteps
26      BWsmooth   = imerode(BWsmooth,seD);
27  end
28  BW=BWsmooth;      %Output
```

D.2 Angular resolved boundary positions of cell clusters

With the cluster's center of mass, calculated from the first frame of the image stack, the distance between center and boundary is measured along angular resolved segments. For circular structures and the outside of donut-shaped structures, the boundaries were detected while approaching the region of interest from the borders of the image, in contrast the inside boundary of donuts was detected by approaching the region of interest from the center.

Code for approaching the boundaries from the image borders (circles and outside of donuts)

```
1   function [dangles,dists, edges] = ...
        angular_distance(L,centroid,angle_incr)
2   % calculation of the disance (in pixel units) from image
3   % border to first outline (circular or donut shaped patterns)
4   % L needs to be a labeled ROI (bwlabel), class: double
5   % Angle increment must be 0<angle_incr<90
6   % top:0deg, right:90deg, bottom: 180deg, left:270deg
7
8   lastangle=360-angle_incr;
9   dangles=[0: angle_incr:lastangle]';
10      %do not count 0 and 360 deg twice
11
12  v=L(round(centroid(1,2)),round(centroid(1,1)));
13      % v is centroid of first frame
14
15  k=1;
16  for i=0:angle_incr:lastangle
17      angle=2*pi*(i+90)/360;
18          %+90 is only necessary that 0 deg is on the top
```

```
        %%scan from center to outside and detect image borders
        j=0; l=1;
        while l == 1
            j=j+1;
            px=j*-cos(angle)+centroid(1,1);
                %- for clockwise turning direction
            py=j*-sin(angle)+centroid(1,2);
                %- for the starting point on the top
            pxi=round(px);
            pyi=round(py);
            if pxi == 1 | pxi == size(L, 2) | pyi == 1 | pyi == ...
                size(L,1)
                l=0;
            end
        end

        o=L(pyi,pxi);
        po=o;
        while po == o
            j=j-1;
            px=j*-cos(angle)+centroid(1,1);
            py=j*-sin(angle)+centroid(1,2);
            pxi=round(px);
            pyi=round(py);
            po=L(pyi, pxi);
        end

        %% increase the accuracy of the measurement
        j=j+1; jj=0;po=o;
        while po == o
            jj=jj+1;
            l=j-jj*0.1;
            px=l*-cos(angle)+centroid(1,1);
            py=l*-sin(angle)+centroid(1,2);
            pxi=round(px);
            pyi=round(py);
            po=L(pyi, pxi);
        end
        dists(k,:)=sqrt((pxi-centroid(1,2))^2+(pyi-centroid(1,2))^2);
        edges(k,:)=[i pxi pyi];
        k=k+1;
end
```

Code for approaching the boundaries from the object's center (inside of donuts)

```
1  function [dangles,distsChild, edgesChild] = ...
       angular_distance(L,centroid,angle_incr)
2  % calculation of the disance (in pixel units) from center to
3  % first boundary of a donut-shaped ROI
4  % L needs to be a labeled ROI (bwlabel), class: double
5  % Angle increment must be 0<angle_incr<90
6  % top:0 deg, right: 90 deg, bottom: 180 deg, left:270 deg
7
8  lastangle=360-angle_incr;
9  dangles=[0: angle_incr:lastangle]';
10        %do not count 0 and 360 deg twice
11 v=L(round(centroid(1,2)),round(centroid(1,1)));
12        % v is centroid of first frame
13
14 k=1;
15 for i=0:angle_incr:lastangle
16     angle=2*pi*(i+90)/360;
17         %+90 makes 0 deg on the top
18
19     %% scan from center towards outside
20     j=0; pv = v;
21     while pv == v
22         j=j+1;
23         px=j*-cos(angle)+centroid(1,1);
24             %- for clockwise turning direction
25         py=j*-sin(angle)+centroid(1,2);
26             %- for the starting point on the top
27         pxi=round(px);
28         pyi=round(py);
29         pv=L(pyi, pxi);
30     end
31
32     %% increase the accuracy
33     j=j-1; jj=0; pv=v;
34     while pv == v
35         jj=jj+1;
36         l=j+jj*0.1;
37         px=l*-cos(angle)+centroid(1,1);
38         py=l*-sin(angle)+centroid(1,2);
39         pxi=round(px);
40         pyi=round(py);
41         pv=L(pyi, pxi);
42     end
```

```
43      distsChild(k,:)=sqrt((pxi-centroid(1,2))^2+ ...
            (pyi-centroid(1,2))^2);
44      edgesChild (k,:)=[i pxi pyi];
45      k=k+1;
46  end
```

Appendix E

List of Publications

Publications

Rolli CG, Seufferlein T, Kemkemer R, Spatz JP (2010) Impact of tumor cell cytoskeleton organization on invasiveness and migration: a microchannel-based approach. *PloS one* 5:e8726 doi:10.1371/journal.pone.0008726.

Rolli CG, Nakayama H, Yamaguchi K, Spatz JP, Kemkemer R, Nakanishi J; Switchable adhesive substrates: Revealing geometry dependence in collective cell behavior *submitted*.

Bai AHC, Milde T, Remke M, **Rolli CG**, Hielscher T, Cho Y-J, Kool M, Northcott PA, Jugold M, Bahzin AV, Eichmüller SB, Kulozik AE, Pscherer A, Benner A, Taylor MD, Pomeroy SL, Kemkemer R, Witt O, Korshunov A, Lichter P, Pfister SM; MicroRNA-182 promotes metastatic dissemination of non-sonic hedgehog-medulloblastoma *Acta Neuropathologica, under review*.

Conference talks

Invited talk at the Gordon Research Conference "Gradient Sensing & Directed Cell Migration" 5-10th June 2011, Les Diablerets, Switzerland: "Photoswitchable wound healing assay to study collective cell migration and leader cell formation".

Acknowledgement

I am grateful to Prof. Joachim P. Spatz for providing me the opportunity to work in a highly dynamic and interdisciplinary research environment. With his support and confidence in my work, I was able to freely plan and conduct the experiments, spend a research stay in Japan and present my work on national and international conferences.

Prof. Michael Grunze kindly accepted to review my thesis and I thank him for dealing with this interdisciplinary subject. Prof. Harald Herrmann, who is also in the evaluation committee, has enriched my view on biological systems during several discussions.

In particular, I am indebted to Ralf Kemkemer, my supervisor, for the stimulating (scientific and non-scientific) discussions, his strong support and the great working atmosphere I experienced during the whole time of my doctorate. I am thankful to our collaboration partner Jun Nakanishi; he frankly received me in Japan, introduced me to the Japanese working culture including the art of nomination and made my stay as pleasant as it was.

I thank my collaboration partners Stefan Pfister and Alfa Bai at the German Cancer Research Center (DKFZ) and Prof. Thomas Seufferlein (University of Halle) for stimulating and enriching discussions.

My office mates, fellow students and colleagues Borja Aragüés, Melih Kalafat, Jovana Matic, Martin Deibler, Christian Eberhard, Stefan Kudera, Kristen Mills, Tobias Busch, Lindarti Purwaningsih and Tuna Ötztürk were always there when a suggestion was needed, a problem to be resolved, a weekend to be spent together or just a beer to be drunk. Over the years, also Mercedes Dragovits, Ada Cavalcanti-Adam, Vera Hirschfeld-Warnecken and Claudia Pacholski were good friends and provided valuable advice.

I appreciate the support of Elisabeth Pfeilmeier und Jutta Hess who helped me to solve administrative challenges. I thank for the support of every member of our technical staff, who were always helpful when needed, especially Jennifer Diemer who became an expert in migration chip assembly within the last year and Christine Mollenhauer with her exhilarating character was always there to cheer me up in the lab.

Also a big thank you to the whole Spatz group with its members in Heidelberg and Stuttgart for the good atmosphere, valuable discussions and help I received. In Japan, Hidekazu Nakayama and Shingo Kaneko were great colleagues and Hiroko Komura was of indispensable value for managing administrative affairs.

My friends and fellow students from University Timm Fehrentz, Ralph Moser, Christoph Elsässer, Richard Schmidt, Alexander Urich, Diana Vogel and Sophie Föhr, although being scattered over many places, provided encouraging support. I also want to express my gratitude to Masaru Komatsu and his whole family for the unforgettable time we spent together in Kyushu.

Finally, I would like to thank my family for their encouragement, personal backup and for being always there when I needed them.

I want morebooks!

Buy your books fast and straightforward online - at one of the world's fastest growing online book stores! Environmentally sound due to Print-on-Demand technologies.

Buy your books online at

www.get-morebooks.com

Kaufen Sie Ihre Bücher schnell und unkompliziert online – auf einer der am schnellsten wachsenden Buchhandelsplattformen weltweit! Dank Print-On-Demand umwelt- und ressourcenschonend produziert.

Bücher schneller online kaufen

www.morebooks.de

OmniScriptum Marketing DEU GmbH
Heinrich-Böcking-Str. 6-8
D - 66121 Saarbrücken

Telefax: +49 681 93 81 567-9

info@omniscriptum.de
www.omniscriptum.de

Printed by Books on Demand GmbH, Norderstedt / Germany